SpringerBriefs in Applied Sciences
and Technology

Computational Mechanics

For further volumes:
http://www.springer.com/series/8886

Ramón Quiza · Omar López-Armas
J. Paulo Davim

Hybrid Modeling
and Optimization
of Manufacturing

Combining Artificial Intelligence
and Finite Element Method

 Springer

Ramón Quiza
Department of Mechanical Engineering
University of Matanzas
Autopista a Varadero
44740 Matanzas
Cuba

J. Paulo Davim
Department of Mechanical Engineering
University of Aveiro
Campus Santiago
3810-193 Aveiro
Portugal

Omar López-Armas
Department of Mechanical Engineering
University of Matanzas
Autopista a Varadero, km 3½
44740 Matanzas
Cuba

ISSN 2191-5342
ISBN 978-3-642-28084-9
DOI 10.1007/978-3-642-28085-6
Springer Heidelberg New York Dordrecht London

e-ISSN 2191-5350
e-ISBN 978-3-642-28085-6

Library of Congress Control Number: 2012931420

Printed on acid-free paper

Springer is part of Springer Science+Business Media (www.springer.com)

Preface

Neither the finite element method nor the artificial intelligence are new tools in modeling and optimization of manufacturing processes. A big amount of information on these topics can be found in the specialized literature. Nevertheless, the combination of both approaches in not as old and widespread. Only in recent years, this approach is beginning to receive a noticeable attention from the research community.

Combining the capability of the finite element method for computing good approximate solutions of partial differential equations defined on geometrically complicated domains with the advantages of artificial intelligence-based techniques for mapping nonlinear noisy relationships and for obtaining near-optimal solutions in complex problems, results in a more powerful and flexible tool. This hybrid approach has been used for solving some problems in manufacturing modeling and optimization, but it is currently in its birth. It can be undoubtedly expected that, in the next years, the growing of the processing power of computers together with the development of new more efficient methods in both areas, increase the efficacy and efficiency of this methodology.

The main objective of this text is to expose some conceptual ideas on the integration of the finite element method and artificial intelligence tool for solving modeling and optimization problems in the field of manufacturing processes. Also, the main topics on both tools are explained and an illustrative sample of the use of the hybrid approach is presented.

The book is directed to the research community dedicated to the mathematical modeling and optimization of manufacturing processes. It is intended to be employed at postgraduate level but some of its topics can be used also in undergraduate courses.

As in every work, the contribution of many people played an important role. However, we want to highlight the support of our college Professor Marcelino

Rivas, through all the processes of conceiving and writing of the book. His advises and suggestions were actually invaluable for this work.

R. Quiza
O. López-Armas
J. P. Davim

Contents

1 Introduction 1
  1.1 Relevance and Convenience of Hybrid Modeling
     and Optimization of Manufacturing Processes............ 1
  1.2 Approaches for Combining AI and FEM 4
     1.2.1 FEM/AI Models 4
     1.2.2 AI/FEM Models 5
     1.2.3 Hybrid Approaches for Optimization............... 6
     1.2.4 Fuzzy FEM 8
  References 9

2 Finite Element in Manufacturing Processes.................. 13
  2.1 Basis of the Finite Element Method..................... 13
  2.2 FEM for Linear Elastostatic Problems 16
  2.3 FEM for Plasticity................................... 20
     2.3.1 Plasticity Fundamentals 20
     2.3.2 Material Behavior Models 21
     2.3.3 Yielding Criteria.............................. 24
     2.3.4 Governing Equations.......................... 24
     2.3.5 FEM Formulation 29
  2.4 Thermal Analysis 30
  2.5 Friction Models..................................... 31
  2.6 Fracture 33
  References 36

3 Artificial Intelligence Tools.............................. 39
  3.1 Preliminary Concepts................................ 39
  3.2 Artificial Neural Networks 40
     3.2.1 Biological Foundations and Neuron Model........... 40
     3.2.2 Network Topology and Learning 41
     3.2.3 Multilayer Perceptron 43

	3.2.4	Radial Basis Function Networks	46
	3.2.5	Hopfield Networks	49
	3.2.6	Adaptive Resonance Theory and Self-Organizing Maps	51
	3.2.7	Warnings and Shortcomings in the Use of Neural Networks	56
3.3	Fuzzy Logic		58
3.4	Neuro-Fuzzy Systems		61
3.5	Metaheuristic Optimization		64
	3.5.1	Optimization Basis	64
	3.5.2	Evolutionary Computation	66
	3.5.3	Evolutionary Multi-Objective Optimization	68
	3.5.4	Swarm Intelligence	71
References			76

4	Case of Study		79
4.1	Case Description		79
4.2	Finite Element Method Based Modeling		79
	4.2.1	Model Description	79
	4.2.2	Outcomes of the FEM	80
4.3	Empirical Modeling		85
	4.3.1	Statistical Modeling	85
	4.3.2	Neural Network-Based Modeling	86
4.4	Optimization		89
4.5	Concluding Remarks		91
References			91

| Index | | | 93 |

Chapter 1
Introduction

Abstract This chapter begins with an explanation about the importance of modeling and optimization of manufacturing processes not only from the scientific and researching point of view but also for practical industrial applications. Then it introduces the hybrid approach which combines artificial intelligence tools and finite element method for these modeling and optimization tasks. The advantages and shortcomings of each of these techniques are exposed, highlighting the convenience of combining both methods, increasing the robustness and flexibility. Furthermore, the different approaches for combining artificial intelligence and finite element method in modeling and optimization of manufactured processes are outlined and preliminarily evaluated.

1.1 Relevance and Convenience of Hybrid Modeling and Optimization of Manufacturing Processes

Modeling of the physical phenomena involved in manufacturing processes (machining, forming, foundry, etc.) has been recognized as one of the most important tasks in manufacturing research. Accurate and realistic mathematical models not only allow understanding how these phenomena take place, but also facilitate the development of new manufacturing processes.

Knowledge about the quantitative relationships between the different parameters involved in manufacturing processes also permits the implementation of effective monitoring and control system, which are indispensable in the high automated modern industry.

Moreover, the use of highly optimized manufacturing processes is widely accepted as a necessary condition for achieving effectiveness, efficiency and economic competitiveness in manufacturing workshops.

R. Quiza et al., *Hybrid Modeling and Optimization of Manufacturing*,
SpringerBriefs in Computational Mechanics,
DOI: 10.1007/978-3-642-28085-6_1, © The Author(s) 2012

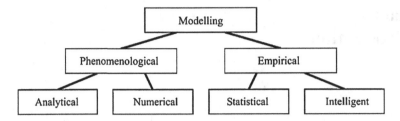

Fig. 1.1 Classification of modeling techniques

Unfortunately, the physical nature of the phenomena underlying the manufacturing processes is not easy to understand, as they involve complex nonlinear relationships which are no completely explained up to date. This situation worsens with the use of modern materials for parts and tools. For example, for many years, the tool life of high speed steels cutting tools, working at relatively low speeds, was described by the well-known Taylor's law (Childs et al. 2000) with an acceptable error level. However, the introduction of multi-coated carbides and PCBN tools, at high cutting speeds and hard conditions has made the Taylor's law useless or, at less, extremely limited (Dolinšek et al. 2001).

Two main approaches can be used in modeling manufacturing processes (see Fig. 1.1). On one hand, the phenomenological modeling is based on the identification and mathematical description of the physical phenomena. It has the advantage of being more realistic and accurate. Also they help to understand the mechanisms of these phenomena. This mathematical description comes in the form of expressions, usually differential equations which can be solve analytically only for a limited set of simple problems. For most of the real problems, these equations (especially partial differential equations) have not analytical solutions and approximated numerical solutions must be obtained instead.

Between the numerical methods used for solving partial differential equation, the finite element method (FEM) has reach the higher application levels, because of its ability for being applied to problems defined over complex spatial domains and the relative simplicity of its computational implementation (Dixit and Dixit 2008). FEM has been successfully applied to a wide variety of manufacturing processes, such as machining (Arrazola and Özel 2010; Mamalis et al. 2008), forming (Gudur and Dixit 2008; Shahani et al. 2009), welding (Anca et al. 2011) and foundry (Lewis et al. 2005).

Sometimes, the phenomenological models are too much idealized for giving results accurate enough for being used in practical applications. In these cases, empirical models, based on correlation of experimental data, play a crucial role. The main drawback of this kind of models is its incapability for identifying or explaining the physical relationship between the involved variables. However, they usually offer accurate outcomes for industrial applications and had been widely used in optimization of manufacturing processes.

Traditionally, the mathematical techniques used for correlating data in empirical models were those based on statistics. These tools are quite simple and are supported by a well-developed theory. Furthermore, there is a lot of computer software allowing the applications of a wide variety of statistical techniques.

Statistical-based empirical models has been successfully used through the time, from the first reported work in manufacturing modeling and optimization (Taylor 1907) to very recent papers in this field (Sahin and Motorcu 2008; Palanisamy et al. 2007).

Nevertheless, the introduction of new materials, such as composites, ceramics and coated carbides) and the use of high work conditions (high cutting speed, in machining, or high deformation rates, in forming, for example) have shown the shortcoming of the statistical techniques. The relationships between different variables, in the new conditions, are commonly too complex for being properly described by a single regression equation.

Some artificial intelligence (AI) techniques, such as artificial neural networks and support vector machines, has proved being effective in matching complex and noisy relationships, and has been widely applied in modeling of manufacturing process (Singha and Gupta 2010; Kolodziejczyk et al. 2010; Wang et al. 2008).

Furthermore, uncertainty has been recognized as taking place in the description of the complex phenomena involved in mechanical manufacturing processes. Here, some AI techniques, such as fuzzy logic, can be a valuable tool (Ching-Kao and Lu 2007; Manabe et al. 2006). Combining neural networks and fuzzy logic in neuro-fuzzy systems, often results in a more powerful and versatile approach, which is also used in modeling of manufacturing processes (Dutta et al. 2006; Coutinho and Marinescu 2005).

On the other hand, AI techniques have application not only in the field of modeling but also in the optimization of manufacturing processes. Due to the complexity of involved relationships, which play the role of objective function and constraints, in the optimization problem, conditions such as continuity, differentiability and unimodality cannot be guaranteed; therefore, the applicability of numeric optimization techniques, in these cases, is often very limited.

On the contrary, heuristic optimization techniques, such as evolutionary algorithms, ant-colony algorithm and simulated annealing have been reported as successful in dealing with this kind of problems (Chandrasekaran et al. 2010).

The main drawback of the application of AI techniques in modeling is the need of large datasets in order to carry out the training process (Quiza and Davim 2011). Furthermore, the theory supporting these approaches is not as well developed as in statistics, sometimes they not are applied with all the care and mathematical rigor they require (Sha and Edwards 2007).

Combining both approach, i.e. FEM and AI-based techniques, can result in a more powerful and flexible approach. Some attempts had been done in this sense, in several branches of the manufacturing processes modeling and optimization (Fu et al. 2010; Chan et al. 2008; Umbrello et al. 2008). In the following pages, some theoretical background on this approach is presented, combined with a literature review. Finally, an application sample is presented in order to illustrate how this hybrid approach works.

1.2 Approaches for Combining AI and FEM

1.2.1 FEM/AI Models

The combination of FEM and AI-based technique can be carried out by following different approaches. The first one (called here FEM/AI) is based on using the FEM to simulate the manufacturing process under different conditions (materials, tools, working parameters, etc.). After that, the output of these simulations is employed as a dataset for training some AI-based model. This approach is represented graphically in Fig. 1.2. The use of the FEM has the key advantage of avoiding the requirement of great amount of experimental data as training set for the AI-based model. On the other hand, the physical laws underlying in the FEM cannot be integrated directly in the AI-based model, because their numerical solution takes too much time.

In this methodology, the FEM-based model is used to simulate some parameters, such as forces or temperatures, depending on different working conditions. This model requires as input not only theoretical assumptions, such as physical laws of the involved phenomena: viscoplasticity, fracture, friction, fluid dynamic, heat transfer, etc., but also experimental data describing part of these phenomena. The main reason for doing that is the complexity of this kind of processes, which have not been fully described, until now, with purely theoretical models.

The output of the FEM-based models forms a simulated data set relating different working parameters and the values of the studied variables. It is used as the training set for the AI-based model, usually a neural network, a fuzzy inference system or a neuro-fuzzy approach.

It must be pointed that this FEM/AI approach has been employed in modeling several manufacturing processes and is the most reported methodology in the literature (Sharma et al. 2009; Shahani et al. 2009; Zhang et al. 2008; Chan et al. 2008; Gudur and Dixit 2008; Umbrello et al. 2008; Umbrello et al. 2007; Manabe et al. 2006; Lin and Lo 2005; Boillat et al. 2004; Ray and MacDonald 2004; Kim and Kim 2000).

For example, in the work of Umbrello et al. (2008), the residual stress, in a hard turning of AISI 52100 bearing steel, is predicted by using a hybrid system composed by a FE-based model and an artificial neural network. The FE-based model uses the rigid plastic constitutive law, Brozzo's fracture criterion and a hardness dependent flow stress behavior to represent the physics and thermodynamics of hard machining in a numeric basis. For every simulated condition values of hardness, chamfered edge, feed rate, cutting speed and chamfer angle were established, and the corresponding parameters describing the axial and hoop residual stress profiles were obtained.

After that, the obtained dataset was used to train a three layer perceptron. In this study, the steepest descent optimization method (simple backpropagation) was used together with Bayesian regularization in training the neural networks. After the training process, the neural network showed a good generalization capability, being able to predict the profile of the residual stress under different conditions.

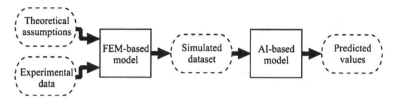

Fig. 1.2 FEM/AI model

In other example, Sharma et al. (2009) used a combination of the FEM and a neuro-fuzzy system for predicting the extrusion load in a hot extrusion process. For different values of die angle, friction coefficient and temperature, the corresponding values of extrusion load were obtained by using the FEM. Then, a neuro-fuzzy inference system was trained with the simulated dataset. Finally, the whole system exhibited a good prediction capability not only for the training set but also for the independent validation set.

An important feature of this approach is the possibility of using different specialized software for implementing the FE-based and AI-based models. For example, DEFORM or ADVANTEDGE FEM can be used for developing the finite element based modeling of machining, and MATLAB or WEKA, for artificial intelligence techniques.

1.2.2 AI/FEM Models

Other approach for combining the FEM and AI techniques is based in the use of AI-based models as input for the FEM. This methodology, called AI/FEM, is graphically represented in Fig. 1.3.

Here, the experimental data is processed by some AI-based tool, such as fuzzy logic or neural networks, in order to identify and generalize the relationship between the empirical dataset.

The obtained empirical model, together with the proper theoretical assumptions, is then introduced to the FEM-based model for predicting some variables of the manufacturing process.

Several interesting papers have been published supporting this approach. Javadi et al. (2003) have reported an intelligent finite element approach where a neural network is incorporated into the FEM code for modeling the constitutive relationship of the material. They claim that the main benefits of using a neural network approach are that all aspects of material behavior can be implemented within a unified environment of a neural network; there is no need for complicated mathematical relationships; there are no material parameters to be identified; and the network is trained directly from experimental data.

Sun et al. (2010) have proposed another neural network-based model. This one predicts the flow stress for a titanium alloy, as a function of the temperature, strain,

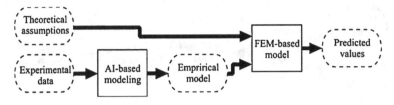

Fig. 1.3 AI/FEM model

and strain rate. In the paper the possibility of using this model in FEM was not noted, but this is implicit. Furthermore, several important issues about the numerical implementation of this approach have been noted (Hashash et al. 2004) to guarantee the convergence of the method.

This approach has been successfully applied to a rolling process (Das et al. 2007) by using a neuro-fuzzy system for predicting material properties, which are used by a cellular automata finite element (CAFE), implemented in ABACUS.

Fu et al. (2010) have presented other interesting approach, using a multilayer perceptron-type neural network for modeling the optimal punch radius, in the air-bending of a sheet metal. With the predicted punch radius and other geometrical parameters, a FEM model is then established, through the use of ABACUS. It must be highlighted in this work the use of a genetic algorithm for optimizing the neural network parameters.

The most relevant aspect of this approach is the possibility of use more flexible and exact models in the FEM-based simulations of machining processes. However, this combination is still very slow in giving the simulated parameters due to it requires solving the FE model each time.

1.2.3 Hybrid Approaches for Optimization

In the two previous sections, approaches combining FEM and AI-based techniques for modeling and simulation of manufacturing processes have been presented. Nevertheless, both tools can be also combined to carry out the optimization of these processes.

As has been pointed out (Chandrasekaran et al. 2010; Quiza et al. 2009), optimization of machining processes is a difficult task, requiring knowledge about the specific manufacturing process, empirical models of input–output and in-process parameters relationship, choosing of realistic constraints, development of convenient optimization criteria, and implementation of effective and efficient optimization techniques.

Moreover, the complex nature of the physical phenomena in the machining processes makes the function relating the different parameters (which are used in the optimization process not only as target functions but also as constraints) do not

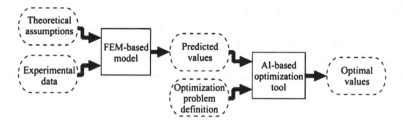

Fig. 1.4 FEM/optimization approach

fulfill the mathematical requirements (such as continuity, differentiability or unimodality) for applying the numerical optimization techniques. On the contrary, some heuristic AI-based tools, into the field of the so-called soft computing, do not demand these prerequisites and has been successfully applied for optimizing manufacturing processes (Chandrasekaran et al. 2010; Mukherjee and Ray 2006).

These heuristic techniques are especially convenient in a posteriori multi-objective optimization problems, where a set of optimal (or near optimal) solutions must be simultaneously determined. In this approach, conventional numerical techniques are remarkably inefficient because of they must be applied several times in order to obtain each solution.

FEM and AI-based optimization tools can be combined as Fig. 1.4 shows. Through the FEM predicted values for the studied variables are obtained, depending on the parameters of the process. These predicted values together with the definition of the optimization process (i.e. the selection of the decision variables, target functions and constraints) are supplied to the AI-based optimization tool in order to obtain the corresponding optimal values.

Certainly, this approach can be enhanced by using AI techniques for modeling the relationship between the predicted values and the process parameters. These models are then used as target functions and constraints for the optimization process.

Some interesting papers can be found in the recent years illustrating the combination of FEM and AI in the optimization of manufacturing processes. Boillat et al. (2004) have modeled a laser sintering process of titanium powder by using a combination of non-standard FEM and finite difference method. After that an invertible model was obtained through a Sugeno-type neuro-fuzzy adaptive inference system. Finally, this model was used in an iterative optimization technique in order to obtain the most convenient density and bounding quality of the manufactured part.

An interesting combination of FEM and genetic algorithm was supplied by Azene et al. (2010) in order to carry out the cooling of a rolling system. In a first step, an FEM-based model predicts the temperature and stress on the roll, for different process parameters. With these values, regression models were fitted. Finally, both objectives (temperature and stress) are simultaneously optimized by using the non-sorting genetic algorithm II (NSGA-II).

Other example of this approach was provided by Khoei et al. (2010). They perform an optimal design for powder die-pressing process based on the genetic algorithm approach. The genetic algorithm is employed to perform an optimal design based on a fixed-length vector of design variables. The technique is used to obtain the desired optimal compacted component by verifying the prescribed constraints.

1.2.4 Fuzzy FEM

Classical FEM was developed for dealing with deterministic problem, where each variable has a defined value. Although sometimes this value is unknown, it can be determined with mathematical precision. Nevertheless, in the real life (and, especially, in the workshop environment), variables deal with some level of non-determinism. This non-determinism can be broadly classified in variability, where the variations are inherent to the modeled physical system (for example, manufacturing tolerances), and uncertainty, which involves a deficiency in the modeling process due to a lack of knowledge (e.g. material constitutive relationships in elastic-viscoplastic behavior) (Verhaeghe et al. 2010). Moreover, these two types of non-determinism are not rigidly separated and often overlap.

Several FEM approaches have been proposed in order to consider the fuzziness of the knowledge involved in practical problems. Probabilistic (or stochastic) FEM was developed for dealing with problems where the respective intervals and probability distributions are known for all the non-deterministic variables. In this method the involved parameters are simulated and the probability values of the outcomes are estimated by carrying out statistical analyses (Muhanna et al. 2006).

Unfortunately, in many real-world applications not only the actual value of the parameters is unknown, but also, there is not available information about the probability distribution of their intervals. For example, the yield strength of a steel grade is constrained by the bounds established in the respective standard. However, the probability distribution depends on the characteristics of the manufacturing process and is very difficult to estimate.

Fuzzy FEM was developed with the main purpose of dealing with data having unknown probability distributions. Fuzzy FEM is based on the classic deterministic FEM. The analysis begins with the definition of the model and the subsequent assembling of the fuzzy system matrices. Concepts of fuzzy arithmetic are then applied, in the second part of the analysis, for computing the outcomes from these fuzzy system matrices (Moens and Vandepitte 2005).

In spite of the proved capability of the fuzzy FEM for successfully dealing with uncertainty in engineering problems, to the best of our knowledge, there are not reported applications of this technique in the field of modeling of manufacturing processes. However, it can be expected that, in near future, this method will be introduced for modeling complex manufacturing processes, allowing considering more realistic conditions and, thus, providing more accurate outcomes.

References

A. Anca, A. Cardona, J. Risso, V.D. Fachinotti, Finite element modeling of welding processes. Appl. Math. Model. **35**, 688–707 (2011). doi:10.1016/j.apm.2010.07.026

P.J. Arrazola, T. Özel, Investigations on the effects of friction modeling in finite element simulation of machining. Int. J. Mech. Sci. **52**, 31–42 (2010). doi:10.1016/j.ijmecsci.2009.10.001

Y.T. Azene, R. Roy, D. Farrugia, C. Onisa, J. Mehnen, H. Trautmann, Work roll cooling system design optimisation in presence of uncertainty and constrains. CIRP J. Manuf. Sci. Tech. **2**, 290–298 (2010). doi:10.1016/j.cirpj.2010.06.001

E. Boillat, S. Kosolov, R. Glardon, M. Loher, D. Saladin, G. Levy, Finite element and neural network models for process optimization in selective laser sintering. P I Mech. Eng. B.-J. Eng. **218**, 607–614 (2004). doi:10.1243/0954405041167121

M. Chandrasekaran, M. Muralidhar, C. Murali Krishna, U.S. Dixit, Application of soft computing techniques in machining performance prediction and optimization: a literature review. Int. J. Adv. Manuf. Tech. **46**, 445–464 (2010). doi:10.1007/s00170-009-2104-x

W.L. Chan, M.W. Fu, J. Lu, An integrated FEM and ANN methodology for metal-formed product design. Eng. Appl. Artif. Intell. **21**, 1170–1181 (2008). doi:10.1016/j.engappai.2008.04.001

T. Childs, K. Mekawa, T. Obiwaka, Y. Yamano, *Metal Machining Theory and Applications* (Wiley, New York, 2000)

C. Ching-Kao, H.S. Lu, The optimal cutting-parameter selection of heavy cutting process in side milling for SUS304 stainless steel. Int. J. Adv. Manuf. Tech. **34**, 440–447 (2007). doi:10.1007/s00170-006-0630-3

R. Coutinho, I. Marinescu, Methodology to compare 3-D and 2-D parameters for the optimization of hard turned surfaces. Mach. Sci. Technol. **9**, 383–409 (2005). doi:10.1080/10910340500196330

S. Das, M.F. Abbod, Q. Zhu, E.J. Palmiere, I.C. Howard, D.A. Linkens, A combined neuro fuzzy-cellular automata based material model for finite element simulation of plane strain compression. Comp. Mater. Sci. **40**, 366–375 (2007). doi:10.1016/j.commatsci.2007.01.010

P.M. Dixit, U.S. Dixit, *Modeling of Metal Forming and Machining Processes by Finite Element and Soft Computing Methods* (Springer, London, 2008)

S. Dolinšek, B. Šuštaršic, J. Kopac, Wear mechanisms of cutting tools in high-speed cutting processes. Wear **250**, 349–356 (2001). doi:10.1016/S0043-1648(01)00620-2

R.K. Dutta, S. Paul, A.B. Chattopadhyay, The efficacy of back propagation neural net-work with delta bar delta learning in predicting the wear of carbide inserts in face milling. Int. J. Adv. Manuf. Tech. **31**, 434–442 (2006). doi:10.1007/s00170-005-0230-7

P.P. Gudur, U.S. Dixit, A neural network-assisted finite element analysis of cold flat rolling. Eng. Appl. Artif. Intell. **21**, 43–52 (2008). doi:10.1016/j.engappai.2006.10.001

Z. Fu, J. Mo, L. Chen, W. Chen, Using genetic algorithm-back propagation neural network prediction and finite-element model simulation to optimize the process of multiple-step incremental air-bending forming of sheet metal. Mater. Des. **31**, 267–277 (2010). doi:10.1016/j.matdes.2009.06.019

Y.M.A. Hashash, S. Jung, J. Ghaboussi, Numerical implementation of a neural network based material model in finite element analysis. Int. J. Numer. Method Eng. **59**, 989–1005 (2004). doi:10.1002/nme.905

A.R. Khoei, S. Keshavarz, S.O. Biabanaki, Optimal design of powder compaction processes via genetic algorithm technique. Finite Elem. Anal. Des. **46**, 843–861 (2010). doi:10.1016/j.finel.2010.05.004

D.J. Kim, B.M. Kim, Application of neural network and FEM for metal forming processes. Int. J. Mach. Tool Manuf. **40**, 911–925 (2000). doi:10.1016/S0890-6955(99)00090-5

T. Kolodziejczyk, R. Toscano, S. Fouvry, G. Morales-Espejel, Artificial intelligence as efficient technique for ball bearing fretting wear damage prediction. Wear **268**, 309–315 (2010). doi:10.1016/j.wear.2009.08.016

S. Kumar Singha, A. Kumar Gupta, Application of support vector regression in predicting thickness strains in hydro-mechanical deep drawing and comparison with ANN and FEM. CIRP J. Manuf. Sci. Technol. **3**, 66–72 (2010). doi:10.1016/j.cirpj.2010.07.005

A.A. Javadi, T.P. Tan, M. Zhang, Neural network for constitutive modelling in finite element analysis. J. Comput. Assist. Mech. Eng. Sci. **10**, 523–529 (2003)

R.W. Lewis, A.S. Usmani, J.T. Cross, Efficient mould filling simulation in castings by an explicit finite element method. Int. J. Numer. Method Fl **20**, 493–506 (2005). doi:10.1002/fld.1650200606

Y.-Y. Lin, S.-P. Lo, Modeling of chemical mechanical polishing process using FEM and abductive network. Eng. Appl. Artif. Intell. **18**, 373–381 (2005). doi:10.1016/j.engappai.2004.09.008

A. Mamalis, J. Kundrák, A. Markopoulos, D. Manolakos, On the finite element modelling of high speed hard turning. Int. J. Avd. Manuf. Tech. **38**, 441–446 (2008). doi:10.1007/s00170-007-1114-9

K. Manabe, M. Suetake, H. Koyama, M. Yang, Hydroforming process optimization of aluminum alloy tube using intelligent control technique. Int. J. Mach. Tool Manuf. **46**, 1207–1211 (2006). doi:10.1016/j.ijmachtools.2006.01.028

D. Moens, D. Vandepitte, A fuzzy finite element procedure for the calculation of uncertain frequency-response functions of damped structures: Part 1-Procedure. J. Sound Vib. **288**, 431–462 (2005). doi:10.1016/j.jsv.2005.07.001

R. Muhanna, V. Kreinovich, P. Solín, J. Chessa, R. Araiza, G. Xiang, Interval finite element methods: New directions, in *NSF Workshop o Modeling Errors and Uncertainty in Engineering Computations*, Savannah, 2006

I. Mukherjee, P.K. Ray, A review of optimization techniques in metal cutting processes. Comput. Ind. Eng. **50**, 15–34 (2006). doi:10.1016/j.cie.2005.10.001

P. Palanisamy, I. Rajendran, S. Shanmugasundaram, Prediction of tool wear using regression and ANN models in end-milling operation. Int. J. Adv. Manuf. Tech. **37**, 29–41 (2007). doi:10.1007/s00170-007-0948-5

R. Quiza, J.P. Davim, Computational methods and optimization, in *Machining of hard materials*, ed. by J.P. Davim (Springer, London, 2011)

R. Quiza, J.E. Albelo, J.P. Davim, Multi-objective optimisation of multipass turning by using a genetic algorithm. Int. J. Mater. Prod. Tech. **35**, 134–144 (2009). doi:10.1504/IJMPT.2009.025223

P. Ray, B.J. MacDonald, Determination of the optimal load path for tube hydroforming processes using a fuzzy load control algorithm and finite element analysis. Finite Elem. Anal. Des. **41**, 173–192 (2004). doi:10.1016/j.finel.2004.03.005

Y. Sahin, A.R. Motorcu, Surface roughness model in machining hardened steel with cubic boron nitride cutting tool. Int. J. Refract. Met. **H26**, 84–90 (2008). doi:10.1016/j.ijrmhm.2007.02.005

W. Sha, K.L. Edwards, The use of artificial neural networks in materials science based research. Mater. Des. **28**, 1747–1752 (2007). doi:10.1016/j.matdes.2007.02.009

A.R. Shahani, S. Setayeshi, S.A. Nodamaie, M.A. Asadi, S. Rezaie, Prediction of influence parameters on the hot rolling process using finite element method and neural network. J. Mater. Process. Tech. **209**, 1920–1935 (2009). doi:10.1016/j.jmatprotec.2008.04.055

R.S. Sharma, V. Upadhyay, K.H. Raj, Neuro-fuzzy modeling of hot extrusion process. Indian J. Eng. Mater. Sci. **16**, 86–92 (2009)

Y. Sun, W.D. Zeng, Y.Q. Zhao, Y.L. Qi, X. Ma, Y.F. Han, Development of constitutive relationship model of Ti600 alloy using artificial neural network. Comp. Mater. Sci. **48**, 686–691 (2010). doi:10.1016/j.commatsci.2010.03.007

F.W. Taylor, On the art of cutting metals. Trans. ASME **28**, 310–350 (1907)

D. Umbrello, G. Ambrogio, L. Filice, R. Shivpuri, A hybrid finite element method-artificial neural network approach for predicting residual stresses and the optimal cutting conditions during hard turning of AISI 52100 bearing steel. Mater. Des. **29**, 873–883 (2008). doi:10.1016/j.matdes.2007.03.004

D. Umbrello, G. Ambrogio, L. Filice, R. Shivpuri, An ANN approach for predicting subsurface residual stresses and the desired cutting conditions during hard turning. J. Mater. Process. Tech. **189**, 143–152 (2007). doi:10.1016/j.jmatprotec.2007.01.016

W. Verhaeghe, M. De Munck, W. Desmet, D. Vandepitte, D. Moens, A fuzzy finite element analysis technique for structural static analysis based on interval fields, in *4th In-ternational Workshop on Reliable Engineering Computing*, Singapore, pp. 117–128, 2010

X. Wang, W. Wang, Y. Huang, N. Nguyen, K. Krishnakumar, Design of neural net-work-based estimator for tool wear modeling in hard turning. J. Intell. Manuf. **19**, 383–396 (2008). doi:10.1007/s10845-008-0090-8

Y. Zhang, S. Zhao, Z. Zhang, Optimization for the forming process parameters of thin-walled valve shell. Thin Wall Struct. **46**, 371–379 (2008). doi:10.1016/j.tws.2007.10.007

Chapter 2
Finite Element in Manufacturing Processes

Abstract This chapter explains the basis of the finite element method, highlighting the application for manufacturing modeling problems. A review of the principles of plasticity, as used in modeling of machining and forming processes is presented, including the most frequently used constitutive models. The key issues of the finite element method modeling of these mechanical processes are also explained according with the last researches in this field.

2.1 Basis of the Finite Element Method

The finite element method (FEM) has gained popularity in the last years as a powerful numeric method for finding good approximate solutions for systems of partial differential equations. This method is especially suitable when the problem is defined over geometrically complex spatial domains. For this reason, the FEM has been successfully applied to a wide field of engineering problems, such as mechanics of materials (elastic and non-elastic), fluid dynamics, heat transfer and electromagnetism.

The FEM aims to solve a differential equation set:

$$\mathbf{A}(\mathbf{u}) = [A_1(\mathbf{u}), \ A_2(\mathbf{u}), \ \ldots]^{\mathrm{T}} = \mathbf{0}; \qquad (2.1a)$$

in a domain Ω (see Fig. 2.1), being $\mathbf{u}$ the unknown state variable; together with the boundary conditions:

$$\mathbf{B}(\mathbf{u}) = [B_1(\mathbf{u}), \ B_2(\mathbf{u}), \ \ldots]^{\mathrm{T}} = \mathbf{0}; \qquad (2.1b)$$

on the boundary, Γ, of the domain. In these equations, $A_i(\bullet)$ and $B_i(\bullet)$ are differential operators.

R. Quiza et al., *Hybrid Modeling and Optimization of Manufacturing*,
SpringerBriefs in Computational Mechanics,
DOI: 10.1007/978-3-642-28085-6_2, © The Author(s) 2012

Fig. 2.1 Domain and
boundary of a problem

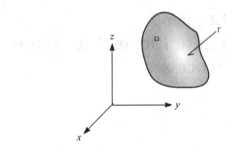

Fig. 2.2 Element domain
and boundary

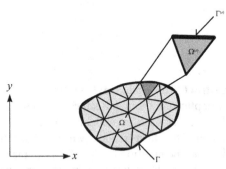

The central idea of the FEM is to replace the exact solution, **u**, by an
approximation, **u***, of the form:

$$\mathbf{u} \approx \mathbf{u}^* = \sum_{i=1}^{n} \mathbf{N}_i \mathbf{a}_i = \mathbf{Na}; \qquad (2.2)$$

where $\mathbf{N}_i$ are the shape functions, predefined in terms of the independent variables
(usually, the coordinates, **x**) and $\mathbf{a}_i$ are parameters, initially unknown, which
should be determined as a result of the application of the method.

In order to obtain this solution, the Eqs. (2.1a, b) must be combined in the so-
called weak form:

$$\int_{\Omega} \mathbf{G}_j(\mathbf{u}^*)d\Omega + \int_{\Gamma} \mathbf{g}_j(\mathbf{u}^*)d\Gamma = \mathbf{0} \qquad j = 1 \ldots n; \qquad (2.3)$$

which permits the approximation to be obtained for every portion of the domain
and assembled (Fig. 2.2):

$$\sum_{e=1}^{m} \int_{\Omega^{(e)}} \mathbf{G}_j(\mathbf{u}^*)d\Omega + \int_{\Gamma^{(e)}} \mathbf{g}_j(\mathbf{u}^*)d\Gamma = \mathbf{0} \qquad j = 1 \ldots n; \qquad (2.4)$$

where $\Omega^{(e)}$ is the domain of the eth portion and $\Gamma^{(e)}$ its part of the boundary
(Zienkiewicz and Taylor 2000).

These portions are known as elements and usually have a simple geometric
shape. Depending on the domain, there are elements with different dimensionality.
For example, a bar (Fig. 2.3a) is a typical one-dimensional element; triangles

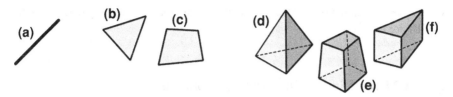

Fig. 2.3 Types of elements **a** Bar element. **b** Triangular element. **c** Quadrangular element. **d** Tetrahedral element. **e** Hexahedral (brick) element. **f** Pentahedral (wedge) element

(Fig. 2.3b) and quadrilaterals (Fig. 2.3c) are the most common two-dimensional elements; and, finally, tetrahedrons (Fig. 2.3d), hexahedrons (Fig. 2.3e) and wedges (Fig. 2.3f) are widely used for meshing three-dimensional domains.

If the differential equations are linear, that is, if the Eq. (2.1a, b) can be written in the form:

$$A(u) = Lu + p = 0 \quad \text{in } \Omega,$$
$$B(u) = Mu + q = 0 \quad \text{on } \Gamma; \tag{2.5}$$

then, the approximating equation system (2.4) yields a set of linear algebraic equations of the form:

$$Ka + f = 0; \tag{2.6}$$

with

$$K_{ij} = \sum_{e=1}^{m} K_{ij}^{(e)} \quad \text{and} \quad f_{ij} = \sum_{e=1}^{m} f_{ij}^{(e)}; \tag{2.7}$$

which can be numerically solved.

There are two main approaches for obtaining the weak formulation in the FEM; they are the functional variational principle and the weighted residual method.

The essence of the variational method is to calculate the total potential, Π, also known as the functional of the system and, then, to consider the stationarity of this total potential:

$$\delta\Pi = 0; \tag{2.8}$$

as an equilibrium condition (Bathe 1996).

On the other hand, the weighted residual method is based on considering that from the (Eq. 2.1a, b) it follows that:

$$\int_{\Omega} v_j^T A(Na) \, d\Omega + \int_{\Gamma} w_j^T B(Na) \, d\Gamma = 0, \quad j = 1 \ldots n; \tag{2.9}$$

where $A(Na)$ and $B(Na)$ represent the residual errors of replace the approximate solution in the differential equation set and in the boundary conditions, respectively, and v_j and w_j are some weighting functions. In the Galerkin method,

$v_j = w_j = N_j$, i.e., the original shape functions are used as weighting (Zienkiewicz and Taylor 2000).

2.2 FEM for Linear Elastostatic Problems

The basic application of the FEM in structural mechanics is in linear elastostatic problems, where on the domain, Ω, there are three unknown fields: the displacement field, $\mathbf{u} = [u_x, u_y, u_z]^T$; the strains field, $\boldsymbol{\varepsilon} = [\varepsilon_{xx}, \varepsilon_{yy}, \varepsilon_{zz}, \varepsilon_{yz}, \varepsilon_{xz}, \varepsilon_{xy}]^T$; and the stress field, $\boldsymbol{\sigma} = [\sigma_{xx}, \sigma_{yy}, \sigma_{zz}, \sigma_{yz}, \sigma_{xz}, \sigma_{xy}]^T$. As result of the load conditions of the domain, the body force, $\mathbf{b} = [b_x, b_y, b_z]^T$, is known on the entire domain. Moreover, at some portion, Γ_u, of the boundary, the values of the displacements are prescribed as equal to $[\mathbf{u}]$, and, at other portion, Γ_t, the values of the tractions are also prescribed as equal to $[\mathbf{t}]$ (see Fig. 2.4). These relationships are known as boundary conditions. The boundary portions must fulfill the conditions:

$$\Gamma_u \cup \Gamma_t = \Gamma \quad \text{and} \quad \Gamma_u \cap \Gamma_t = \varnothing \tag{2.10}$$

A set of equations establishes the relationships between the different variables defined for the problem. The first one is the cinematic equation, which relates the displacements and strains in the entire domain:

$$\boldsymbol{\varepsilon} = \nabla_s \mathbf{u}; \tag{2.11a}$$

where ∇_s represents the symmetric matrix gradient operator:

$$\nabla_s = \begin{bmatrix} \partial/\partial x & 0 & 0 & 0 & \partial/\partial z & \partial/\partial y \\ 0 & \partial/\partial y & 0 & \partial/\partial z & 0 & \partial/\partial x \\ 0 & 0 & \partial/\partial z & \partial/\partial y & \partial/\partial x & 0 \end{bmatrix}^T \tag{2.11b}$$

On the other hand, the equilibrium equation:

$$\nabla_s^T \boldsymbol{\sigma} + \rho \mathbf{b} = \mathbf{0}; \tag{2.12}$$

establishes the relationships between the body force and the stress field, and it is the application of the principle of conservation of lineal momentum. Finally, the constitutive equation relates the strain and stress fields. This is particular of every material at every specific condition. For linear elastic materials, this constitutive equation is given by the generalized Hooke's law:

$$\boldsymbol{\sigma} = \mathbf{C}^E \boldsymbol{\varepsilon}; \tag{2.13a}$$

where $\mathbf{C}^E$ is the tensor of elasticity, which can be written in matrix form as:

Fig. 2.4 Domain and
boundary definitions

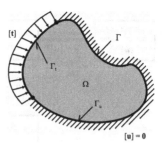

$$\mathbf{C}^E = \begin{bmatrix} c_{11} & c_{12} & c_{13} & c_{14} & c_{15} & c_{16} \\ & c_{22} & c_{23} & c_{24} & c_{25} & c_{26} \\ & & c_{33} & c_{34} & c_{35} & c_{36} \\ & & & c_{44} & c_{45} & c_{46} \\ & \text{sym.} & & & c_{55} & c_{56} \\ & & & & & c_{66} \end{bmatrix} \qquad (2.13b)$$

In the special case of homogeneous and isotropic materials, the matrix of elasticity can be reduced to:

$$\mathbf{C}^E = \frac{E}{(1-2v)(1+v)} \begin{bmatrix} 1-v & v & v & 0 & 0 & 0 \\ & 1-v & v & 0 & 0 & 0 \\ & & 1-v & 0 & 0 & 0 \\ & & & 1-2v & 0 & 0 \\ & \text{sym.} & & & 1-2v & 0 \\ & & & & & 1-2v \end{bmatrix};$$

$$(2.13c)$$

being E, the Young's modulus and v, the Poisson's ratio of the material.

Additionally, from the definition of stress tensor is obtained the relationship that links the stress fields and the prescribed tractions, on the portion of the boundary, Γ_u, where these tractions act:

$$\mathbf{n}\sigma = [\mathbf{t}]; \qquad (2.14a)$$

where

$$\mathbf{n} = \begin{bmatrix} n_x & 0 & 0 & 0 & n_z & n_y \\ 0 & n_y & 0 & n_z & 0 & n_x \\ 0 & 0 & n_z & n_y & n_x & 0 \end{bmatrix} \qquad (2.14b)$$

and n_x, n_y and n_z are the components of the outward normal on the boundary.

This set of equations form the so-called strong formulation of the linear elastostatics, and can be represented, in a very convenient way by using the popular Tonti diagram, as shown in Fig. 2.5.

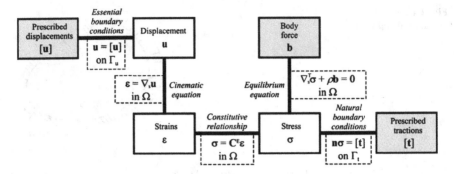

Fig. 2.5 Tonti diagram for the strong form of linear elastostatics

In order to solve this kind of problem by applying the FEM, the strong formulation is replaced by the weak formulation. This enforces the relationships in an integral sense rather than point by point. In elastostatics applications, this weak formulation is given by the Hamilton's principle:

$$\delta \int_{t_1}^{t_2} L \, dt = 0;$$

(2.15)

where the Lagrangian functional, L, is computed as:

$$L = T - U + W;$$

(2.16)

and the kinetic energy, T, the internal energy (here, the elastic strain energy), U, and the work done by the external forces, W, can be defined in the integral forms (Liu and Quek 2003):

$$T = \frac{1}{2} \int_{\Omega} \rho \dot{\mathbf{u}}^T \dot{\mathbf{u}} \, d\Omega;$$

(2.17a)

$$U = \frac{1}{2} \int_{\Omega} \boldsymbol{\varepsilon}^T \boldsymbol{\sigma} \, d\Omega = \frac{1}{2} \int_{\Omega} \boldsymbol{\varepsilon}^T \mathbf{C}^E \boldsymbol{\varepsilon} \, d\Omega;$$

(2.17b)

$$W = \int_{\Omega} \mathbf{u}^T \mathbf{b} \, d\Omega + \int_{\Gamma_t} \mathbf{u}^T \mathbf{t} \, d\Gamma.$$

(2.17c)

As the problem is static, the Hamilton principle can be written as:

$$\delta \left(\int_{\Omega} \mathbf{u}^T \mathbf{b} \, d\Omega + \int_{\Gamma_t} \mathbf{u}^T \mathbf{t} \, d\Gamma - \frac{1}{2} \int_{\Omega} \boldsymbol{\varepsilon}^T \mathbf{C}^E \boldsymbol{\varepsilon} \, d\Omega \right) = 0.$$

(2.18)

As it remains being valid for every element in the discretization, Eq. (2.20) can be rewritten in the form:

$$\delta \left(\int_{\Omega^{(e)}} \mathbf{u}^T \mathbf{b} \, d\Omega + \int_{\Gamma_t^{(e)}} \mathbf{u}^T \mathbf{t} \, d\Gamma - \frac{1}{2} \int_{\Omega^{(e)}} \boldsymbol{\varepsilon}^T \mathbf{C}^E \boldsymbol{\varepsilon} \, d\Omega \right) = 0.$$

(2.19)

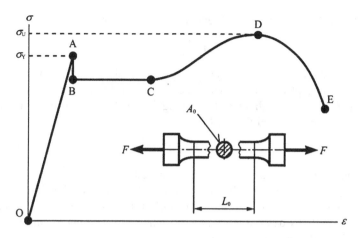

Fig. 2.6 Stress versus strain in a typical tension test

In every element, there are N points (Fig. 2.6), known as nodes, where the displacements, $\mathbf{U}^{(e)} = [u_{x1}, u_{y1}, u_{z1}, \ldots, u_{xN}, u_{yN}, u_{zN}]^{\mathrm{T}}$, can be computed and, then, the displacements in other points, $\mathbf{u}$, can be interpolated from them:

$$\mathbf{u} = \mathbf{N}\mathbf{U}^{(e)}, \tag{2.20}$$

where $\mathbf{N}$ is the matrix of shape functions, depending on the coordinates, $\mathbf{x}$:

$$\mathbf{N} = \begin{bmatrix} N_1(\mathbf{x}) & 0 & 0 & \ldots & N_N(\mathbf{x}) & 0 & 0 \\ 0 & N_1(\mathbf{x}) & 0 & \ldots & 0 & N_N(\mathbf{x}) & 0 \\ 0 & 0 & N_1(\mathbf{x}) & \ldots & 0 & 0 & N_N(\mathbf{x}) \end{bmatrix}$$

By substituting (2.20) in the weak formulation for an element (2.19), and defining the strain-displacement matrix, $\mathbf{B} = \nabla_s \mathbf{N}^{(e)}$, it is obtained the expression:

$$\delta \left(\int_{\Omega^{(e)}} \mathbf{U}^{(e)} \mathbf{N}^{(e)} \mathbf{b}\, d\Omega + \int_{\Gamma_t^{(e)}} \mathbf{U}^{(e)} \mathbf{N}^{(e)} \mathbf{t}\, d\Gamma - \frac{1}{2} \int_{\Omega^{(e)}} \mathbf{U}^{(e)\mathrm{T}} \mathbf{B}^{\mathrm{T}} \mathbf{C}^{\mathrm{E}} \mathbf{B} \mathbf{U}^{(e)}\, d\Omega \right) = 0;$$

which is transformed, after applying the rules of variational calculus, in:

$$\int_{\Omega^{(e)}} \mathbf{N}^{(e)} \mathbf{b}\, d\Omega + \int_{\Gamma_t^{(e)}} \mathbf{N}^{(e)} \mathbf{t}\, d\Gamma - \left(\frac{1}{2} \int_{\Omega^{(e)}} \mathbf{B}^{\mathrm{T}} \mathbf{C}^{\mathrm{E}} \mathbf{B}\, d\Omega \right) \mathbf{U}^{(e)} = 0. \tag{2.21}$$

If the stiffness matrix, $\mathbf{K}^{(e)}$, and the nodal force vector, $\mathbf{F}^{(e)}$, are defined for the element e, as follows:

$$\mathbf{F}^{(e)} = \int_{\Omega^{(e)}} \mathbf{N}^{(e)} \mathbf{b}\, d\Omega + \int_{\Gamma_t^{(e)}} \mathbf{N}^{(e)} \mathbf{t}\, d\Gamma, \tag{2.22a}$$

$$\mathbf{K}^{(e)} = \left(\frac{1}{2} \int_{\Omega^{(e)}} \mathbf{B}^{\mathrm{T}} \mathbf{C}^{\mathrm{E}} \mathbf{B}\, d\Omega \right); \tag{2.22b}$$

the expression (2.21) can be rewritten as:

$$\mathbf{K}^{(e)}\mathbf{U}^{(e)} = \mathbf{F}^{(e)}. \tag{2.23a}$$

The Eq. (2.23a) for all the elements can be assembled together, by considering the equality of displacement in nodes belonging to different elements, and the action-reaction forces on these nodes, giving a global equation involving all the nodal displacement, $\mathbf{U}$, and forces, $\mathbf{F}$:

$$\mathbf{KU} = \mathbf{F} \tag{2.23b}$$

which is an algebraic equation set of the form (2.6).

By considering the essential boundary conditions at nodes belonging to Γ_u, and natural boundary conditions nodes belonging at Γ_t, the system (2.23b) can be simplified, usually by deleting rows and columns corresponding to null degrees of freedom. Then, the obtained system:

$$\tilde{\mathbf{K}}\tilde{\mathbf{U}} = \tilde{\mathbf{F}}; \tag{2.24a}$$

can be numerically solved in order to determine the nodal displacements:

$$\tilde{\mathbf{U}} = \tilde{\mathbf{K}}^{-1}\tilde{\mathbf{F}}. \tag{2.24b}$$

2.3 FEM for Plasticity

2.3.1 Plasticity Fundamentals

As the most important manufacturing processes, such as machining and forming, involves plastic deformations, modeling the plasticity and solving the obtaining models by using numeric methods play a key role in simulation of these processes.

Plasticity can be broadly classified in two types: rate-independent plasticity, where the strain rate has no influence in the strain–stress relationship, and viscoplaticity, where the strain rate has a non-negligible influence. The first one is a good approximation when low strain rates take place in the deformation process; on the contrary, when the strain rates are high, viscoplasticity offers better results.

The behavior of the materials in rate-independent plasticity can be studied through a standard tension test (see Fig. 2.6), where the values of the true stress, σ:

$$\sigma = \sigma_0(1 + \varepsilon); \tag{2.25}$$

are plotted versus the logarithmic strain, ε:

$$\varepsilon = \ln(1 + e); \tag{2.26}$$

where $\sigma_0 = F/A_0$ is the engineering stress applied in the test and $e = (L-L_0)/L_0$ is the infinitesimal linear strain.

Fig. 2.7 Elastic and plastic
components of the strain

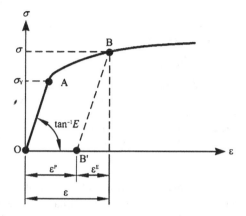

In Fig. 2.6 it is represented a typical experimental stress–strain curve, corresponding to a mild steel (Han and Reddy 1999). Some different regions can be identified in this curve. In portion OA, there is a linear proportionality between the strain
and stress (which is given by the Young's modulus, E), and when the load is retired, the
material return to the unloaded initial condition (point O). This kind of deformation is
known as linear elastic, and follows the previously mentioned Hooke's law (2.6).

When the stress surpasses some value (called yield stress, σ_Y), there is a sharp
sudden drop in the stress value (region AB). The region BC is characterized for a
near zero slope in the curve, that is to say that increments in the strain take place
without any rise in the stress value. The region CD is known as the hardening
region, because the stress increases with the strain, although not with linear
relationship, until achieving the ultimate strength, σ_U, at point D. On the contrary,
in region DE (softening region) increments in strain cause a decrease in the stress
until the final failure at point E.

Even though this curve is representative, actual behavior can strongly change
from one material to another one. Even, heat treatments can change the form of
this curve for the same material.

In spite of the complexity of the material behavior, two well defined zones can
be identified: an elastic region, where deformations disappear after removing the
load, and a plastic region, where some deformations stay after removing the load
(see Fig. 2.7). Therefore, every strain, ε, at the plastic zone can be considered as
composed by an elastic strain, ε^E, and a plastic strain, ε^P:

$$\varepsilon = \varepsilon^E + \varepsilon^P. \tag{2.27}$$

2.3.2 Material Behavior Models

Due to the complexity of the experimental plastic behavior of engineering materials it has been idealized by using simpler models (see Fig. 2.8).

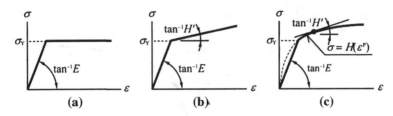

Fig. 2.8 Idealized plastic behavior of materials **a** Elastic-perfectly plastic. **b** Linear work-hardening. **c** Nonlinear work-hardening

In the elastic-perfectly plastic idealization (Fig. 2.8a), there is not further increment in the stress after surpassing the yield point. On the contrary, in the elastic-work-hardening idealization (Fig. 2.8b, c), the stress continues rising with strain increments after the yield point. This increment is modeled by the material hardening function, H:

$$\sigma = H(\varepsilon^{P}). \tag{2.28}$$

For linear work-hardening (Fig. 2.8b), the plastic behavior is represented by a straight line with a constant slope H', while in the nonlinear work-hardening (Fig. 2.8c), the curve changes its slope $H' = dH/d\varepsilon^P$.

Several models have been proposed for the material hardening deformation. The simplest consider that stress is only a function of the strain and not of the strain rate. They are called strain rate-independent plastic models, and include, among others, (Dixit et al. 2011), the Hollomon's law:

$$\sigma = K(\varepsilon^{P})^{n}; \tag{2.29a}$$

that does not fit the stress–strain relationship at low strains; the Ludwik's law:

$$\sigma = \sigma_Y + K(\varepsilon^{P})^{n}; \tag{2.29b}$$

that does not reflect property the constant slope of the stress–strain curve of metals at large strain; the Swift's law:

$$\sigma = \sigma_Y(1 + K\varepsilon^{P})^{n}; \tag{2.29c}$$

which fits better the behavior of the stress–strain curve of metal at large strain; and the Voce's law:

$$\sigma = \sigma_Y + K[1 - m\exp(-n\varepsilon^{P})]; \tag{2.29d}$$

that is more suitable for moderate strain values. In all of these expressions, K and n are experimental constants describing the plastic behavior of the material.

In some circumstances, the effect of the strain-rate cannot be neglected, for example, in cutting processes where high values of strain rates take place. This cases, which include the plastic strain rate term, $\dot{\varepsilon}^P$, in the material hardening function are known as viscoplastic models. If the expression also includes the temperature, T, then is it called thermo-viscoplastic model.

Determination of viscoplastic and thermo-viscoplastic models cannot be carried out through single tension test. It requires more complex methods such as the split Hopkinson pressure bar (Jasper and Dautzenberg 2002). An example of thermo-viscoplastic model is the generalized Oxley's equation (Lalwani et al. 2009):

$$\sigma = \sigma_1(T_{\mathrm{mod}})(\varepsilon^P)^{n(T_{\mathrm{mod}})}; \tag{2.30a}$$

where the coefficient σ_1 and the exponent n are polynomial functions of the strain-modified temperature, T_{mod}:

$$T_{\mathrm{mod}} = T\left(1 - v\log_{10}\frac{\dot{\varepsilon}^P}{\dot{\varepsilon}_0^P}\right); \tag{2.30b}$$

which depends on the material temperature, T, the plastic strain rate, $\dot{\varepsilon}^P$, the reference plastic strain rate, $\dot{\varepsilon}_0^P$, and the strain rate sensitivity constant, v.

The Johnson–Cook's model, also referred as J–C law, is other of the frequently used empirical approaches for modeling the thermo-viscoplastic behavior of materials. It is described by the expression (Özel and Zeren 2004):

$$\sigma = \left[A + B(\varepsilon^P)^n\right]\left[1 + C\ln\left(\frac{\dot{\varepsilon}^P}{\dot{\varepsilon}_0^P}\right)\right]\left[1 - \left(\frac{T - T_0}{T_M - T_0}\right)^m\right]; \tag{2.30c}$$

where the ε is the plastic strain, $\dot{\varepsilon}^P$ is the plastic strain rate, $\dot{\varepsilon}_0$ is the reference strain rate, T is the absolute temperature of the material, T_M is the melting temperature, T_0 is the reference temperature and A, B, C, n and m are material constants (A is the yield strength at T_0, B is the hardening modulus, C is the strain rate sensitivity, n is the strain-hardening exponent, and m the thermal softening exponent). In spite of some limitations with regard to dynamics train aging, i.e. blue-brittleness effect during a certain range of temperature variations in the plastic deformation of carbon steels, the J–C law is very often used to represent the thermo-viscoplastic behavior of workpiece material in manufacturing process modeling, especially in cutting processes (Arrazola and Özel 2010).

Sometimes, the so-called power law (Dixit et al. 2011) is also used for describing the behavior of materials at thermo-viscoplastic state:

$$\sigma = \sigma_0(\varepsilon^P)^n\left(\frac{\dot{\varepsilon}^P}{\dot{\varepsilon}_0^P}\right)^m\left(\frac{T}{T_0}\right)^{-r}; \tag{2.30d}$$

where the terms has the same meaning than in the previous expressions. As in the J-C model, in the power low the effects of strain, strain rate and temperature are considered independent.

Applying the dislocation mechanics theory, Zerilli and Armstrong (Jasper and Dautzenberg 2002) derived other more complex constitutive models for body-centered cubic metals:

$$\sigma = C_0 + C_1 \exp\left(-C_3 T + C_4 T \ln\frac{\dot{\varepsilon}}{\dot{\varepsilon}_0}\right) + C_5 \varepsilon^n; \qquad (2.30\text{e})$$

and for face-centered cubic metal:

$$\sigma = C_0 + C_2 \varepsilon^{1/2} \exp\left(-C_3 T + C_4 T \ln\frac{\dot{\varepsilon}}{\dot{\varepsilon}_0}\right); \qquad (2.30\text{f})$$

where $C_0, C_1, \ldots, C_5$ are material properties. These models has the advantage of a strong theoretical component, however, they have not been as widely applied in FEM-based modeling of manufacturing process as J-C model or Oxley's equation.

2.3.3 Yielding Criteria

Another important aspect, in the theory of plasticity is the initial yielding criterion, that is, the point at which the yield process starts, As follows from Fig. 2.6, the criterion for initial yielding in simple tension is given by:

$$\sigma - \sigma_Y = 0; \qquad (2.31)$$

where σ is the tensile stress and σ_Y, the yield stress of the material. However, more complex stress states require more elaborated criteria. One of the most used criteria for defining the beginning of the yielding process in a material is the Von Mises' criterion, which established that the yield begins when resultant deviatoric stress reaches a critical value. In terms of the principal stresses, σ_1, σ_2 and σ_3, this can be written as:

$$[(\sigma_1 - \sigma_2)^2 + (\sigma_2 - \sigma_3)^2 + (\sigma_3 - \sigma_1)^2] - 2\sigma_Y^2 = 0. \qquad (2.32)$$

On the other hand, the so-called Tresca's criterion, defines the initial yielding from the maximum shear stress, and can be expressed by the equation:

$$[(\sigma_1 - \sigma_2)^2 - \sigma_Y^2][(\sigma_2 - \sigma_3)^2 - \sigma_Y^2][(\sigma_3 - \sigma_1)^2 - \sigma_Y^2] = 0. \qquad (2.33)$$

2.3.4 Governing Equations

Contrary to what happens in elasticity, in plasticity the stress depends on the history of deformation. Mathematically, this can be expressed by using either the incremental form or the rate form (Shabana 2008).

The rate form is used in the so-called Eulerian formulation of the continuum, which considers that the reference coordinate system is fixed and the material moves through it (Fig. 2.9). This approach is very convenient when the material flows through a fixed region of the space, known as control volume, which is used

Fig. 2.9 Eulerian
formulation of the continuum

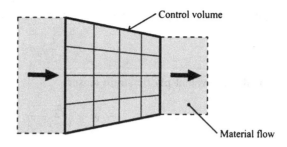

as the problem domain. This Eulerian formulation is frequently applied in extrusion, rolling and other cold forming processes. It is also used in metal cutting, if the shape of the chip is previously known.

There are three main governing equation sets in the Eulerian formulation (Dixit and Dixit 2008). The first one is the kinematic relationship between the velocity vector, v_i, and the strain rate tensor, $\dot{\varepsilon}_{ij}$, which provides six scalar equations[1]:

$$\dot{\varepsilon}_{ij} = \frac{1}{2}\left(\frac{\partial v_i}{\partial x_j} + \frac{\partial v_j}{\partial x_i}\right). \tag{2.34}$$

and the velocity is defined as the rate of change of the material position, x_i:

$$v_i = \frac{dx_i}{dt}. \tag{2.35}$$

The constitutive relation links the elastic–plastic stress and the strain rate and provides other six scalar equations:

$$\dot{\varepsilon}_{kk} = \frac{\dot{s}_{kk}}{3K}; \tag{2.36a}$$

$$\dot{\varepsilon}'_{ij} = \frac{1}{2G}\dot{s}'_{kk} + \frac{3\dot{\varepsilon}^P_{eq}}{2\sigma_{eq}}\sigma'_{ij}; \tag{2.36b}$$

where G and K are the shear modulus and the bulk modulus of the material; $\dot{\varepsilon}'_{ij}$ is the deviatoric part of the strain rate tensor:

$$\dot{\varepsilon}'_{ij} = \dot{\varepsilon}_{ij} - \frac{1}{3}\delta_{ij}\dot{\varepsilon}_{kk}; \tag{2.37}$$

δ_{ij} is the Kronecker delta; σ_{eq} is the Von Mises equivalent stress:

$$\sigma_{eq} = \sqrt{\frac{3}{2}\sigma'_{ij}\sigma'_{ij}}; \tag{2.38}$$

---

[1] For an explanation of the index notation and summation convention used in this section, see Dixit and Dixit (2008, §2.2) or Shabana (2008, §1.3).

σ'_{ij} is the deviatoric part of the stress tensor:

$$\sigma'_{ij} = \sigma_{ij} - \frac{1}{3}\delta_{ij}\sigma_{kk};\tag{2.39}$$

$\dot{\varepsilon}^{P}_{eq}$ is the equivalent plastic strain tensor:

$$\dot{\varepsilon}^{P}_{eq} = \sqrt{\frac{2}{3}\dot{\varepsilon}^{P}_{ij}\dot{\varepsilon}^{P}_{ij}};\tag{2.40}$$

$\dot{s}_{ij}$ is the Jaumann stress rate tensor, which is related with the Cauchy stress rate tensor, $\dot{\sigma}_{ij}$, by the expressions:

$$\dot{s}_{kk} = \dot{\sigma}_{kk} - (\dot{\omega}_{kl}\sigma_{lk} + \sigma_{kl}\dot{\omega}^{T}_{lk});\tag{2.41a}$$

$$\dot{s}'_{ij} = \dot{\sigma}'_{ij} - (\dot{\omega}_{il}\sigma'_{lj} + \sigma'_{il}\dot{\omega}^{T}_{lj});\tag{2.41b}$$

and $\dot{\omega}_{ij}$ is the spin tensor:

$$\dot{\omega}_{ij} = \frac{1}{2}\left(\frac{\partial v_i}{\partial x_j} - \frac{\partial v_j}{\partial x_i}\right).\tag{2.42}$$

The value of the equivalent stress is related with the equivalent strain by the material hardening function:

$$\sigma_{eq} = H(\varepsilon^{P}_{eq});\tag{2.43a}$$

which can also include the strain rate and temperature term, in the general case of thermo-viscoplastic behavior:

$$\sigma_{eq} = H(\varepsilon^{P}_{eq}, \dot{\varepsilon}^{P}_{eq}, T).\tag{2.43b}$$

The third set is given by the motion equation, which consist on three scalar equations:

$$\rho\left(\frac{\partial x_i}{\partial t} + \frac{\partial v_i}{\partial x_j}v_j\right) = \rho b_i + \frac{\partial \sigma_{ij}}{\partial x_j};\tag{2.44}$$

where ρ is the material density and b_i the body force vector.

Additionally, the boundary conditions must be considered. On one hand, the velocity components, $[v]_i$, must be known at some sections of the boundary, Γ_v:

$$v_i = [v]_i, \quad \text{on } \Gamma_v.\tag{2.45a}$$

On the other hand, the values of tension vector, $[t^n]_i$, along the normal to the surface, n_i, must be prescribed on some part of the boundary, Γ_t:

$$\sigma_{ij}n_j = [t^n]_i, \quad \text{on } \Gamma_t.\tag{2.45b}$$

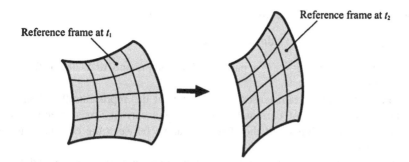

Fig. 2.10 Lagrangian formulation of the continuum

Finally, when dealing with a transient problem, the initial values of the velocities, and hydrostatic and deviatoric par of the stress must be known at every point of the domain, Ω:

$$v_i = v_i^0, \quad \sigma_{kk} = \sigma_{kk}^0, \quad \sigma_{ij}' = \sigma_{ij}'^0, \quad \text{at } t = t_0, \quad \forall x_i \in \Omega. \tag{2.46}$$

In steady-state problems, these initial conditions are not need.

If the behavior of the material is considered as rigid-plastic (i.e., the elastic part of the strain is neglected), an interesting simplification takes place. In this case, the constitutive relation take the form:

$$\dot{\varepsilon}_{kk} = 0; \tag{2.47a}$$

$$\dot{\varepsilon}_{ij}' = \frac{3\dot{\varepsilon}_{eq}^P}{2\sigma_{eq}}\sigma_{ij}'; \tag{2.47b}$$

This change is not trivial because the time derivative of the stress tensor disappears from the constitutive relations (2.47a, b). Thence, although still being nonlinear and, therefore, requiring an iterative solving scheme, the solution of the problem is easier than in the elastic–plastic formulation. Furthermore, only the velocities at the start time are needed as initial conditions.

Nevertheless, the rigid-plastic formulation has some shortcomings related with the neglect of the elastic component of the strain. For example, neither the stress distribution in the non-plastic region nor the residual stresses can be computed by using this approach.

The Lagrangian formulation uses a reference frame which is attached to the material and moves together with it (Fig. 2.10). It is very convenient when the space occupied by the material is not previously known as in free forging or transient cutting analysis.

The Lagrangian approach uses the incremental form which is based on the use of the incremental strain tensor, $\mathrm{d}\varepsilon_{ij}$ (Dixit and Dixit 2008):

$$\mathrm{d}\varepsilon_{ij} = \frac{1}{2}\left[\frac{\partial(\mathrm{d}u_i)}{\partial x_j} + \frac{\partial(\mathrm{d}u_j)}{\partial x_i}\right]; \tag{2.48}$$

defined from the incremental displacement vector, du_i:

$$du_i = x_i(t + dt) - x_i(t); \qquad (2.49)$$

where $x_i(t)$ is the position of a particle at instant t, and dt is the differential increment of time.

In the incremental strain tensor, the component $d\varepsilon_{ii}$ represents the change in current length per unit of current length along the current i-axis direction. The component $d\varepsilon_{ij}$ represents half the change in angle between the current i- and j-axes directions.

Both, the incremental strain tensor and the strain rate tensor are related by:

$$d\varepsilon = \dot{\varepsilon} dt. \qquad (2.50)$$

The constitutive relation links the incremental strain tensor with the increment Jaumann stress tensor, ds_{ij}, or the stress rate tensor. After yielding, i.e., for the elastic–plastic behavior, this constitutive relation can be expressed as:

$$ds_{ij} = C_{ijkl}^{EP} d\varepsilon_{kl}; \qquad (2.51)$$

where C_{ijkl}^{EP} is the fourth order elasticity–plasticity tensor:

$$C_{ijkl}^{EP} = 2G \left[\frac{v}{1 - 2v} \delta_{ij}\delta_{kl} + \delta_{ik}\delta_{jl} - \frac{9G}{2} \frac{\sigma_{ij}'\sigma_{ij}'}{(H' + 3G)\sigma_{eq}^2} \right]; \qquad (2.52)$$

G is the shear modulus; v is the Poisson's ratio; and H', is the slope of the hardening function of the material (Eq. 2.41a, b); σ_{eq} is the equivalent stress (Eq. 2.38), and is σ_{ij}', the deviatoric part of the stress tensor (Eq. 2.39).

Before yielding and after uploading, i.e., when material is behaving elastically, the constitutive equation takes the form:

$$ds_{ij} = C_{ijkl}^{E} d\varepsilon_{kl}; \qquad (2.53)$$

being:

$$C_{ijkl}^{E} = 2G \left(\frac{v}{1 - 2v} \delta_{ij}\delta_{kl} + \delta_{ik}\delta_{jl} \right). \qquad (2.54)$$

The incremental Jaumann stress tensor is related with the incremental Cauchy stress tensor by the equation:

$$ds_{ij} = s_{ij} dt = d\sigma_{ij} - (d\omega_{il}\sigma_{lj} + \sigma_{il}d\omega_{lj}^{T}); \qquad (2.55)$$

where the incremental infinitesimal rotation tensor, $d\omega_{ij}$, is defined as:

$$d\omega_{ij} = \frac{1}{2} \left[d\left(\frac{\partial u_i}{\partial x_j}\right) - d\left(\frac{\partial u_j}{\partial x_i}\right) \right]. \qquad (2.56)$$

The incremental equation of motion states that:

$$\rho d a_i = \rho d b_i + d\left(\frac{\partial \sigma_{ij}}{\partial x_j}\right); \tag{2.57}$$

where db_i is the incremental body force and da_i is the incremental acceleration.

For the Lagrangian formulation, the required boundary conditions are defined by the prescribed incremental displacements, $[du]_i$, at some region Γ_u:

$$du_i = [du]_i, \quad \text{on } \Gamma_u \tag{2.58a}$$

and the prescribed tension vectors $[t^n]_i$ along the normal, n_i, on some region Γ_t:

$$d\sigma_{ij} n_j = [dt^n]_i, \quad \text{on } \Gamma_u. \tag{2.58b}$$

The initial conditions include values of the incremental displacements, du_i, and incremental velocities, dv_i, at time t_0, in every point of the problem domain, Ω:

$$du_i = du_i^0, \ dv_i = dv_i^0, \quad \text{at } t = t_0, \quad \forall x_i \in \Omega. \tag{2.59}$$

2.3.5 FEM Formulation

In the Eulerian finite element formulation, for elastic–plasticity is based in the equation:

$$\mathbf{K}^{(e)} \mathbf{V}^{(e)} = \dot{\mathbf{F}}^{(e)}; \tag{2.60}$$

where $\mathbf{V}^{(e)} = [v_{x1}, v_{y1}, v_{z1}, \ldots, v_{xN}, v_{yN}, v_{zN}]^T$ is the vector of nodal velocities; $\dot{\mathbf{F}}^{(e)} = [\dot{f}_{x1}, \dot{f}_{y1}, \dot{f}_{z1}, \ldots, \dot{f}_{xN}, \dot{f}_{yN}, \dot{f}_{zN},]^T$ is the vector of nodal forces; $\mathbf{K}^{(e)}$ is the element stiffness matrix, given by:

$$\mathbf{K}^{(e)} = \frac{1}{2} \int_{\Omega^{(e)}} \mathbf{B}^T \mathbf{C}^{EP} \mathbf{B} d\Omega; \tag{2.61}$$

and $\mathbf{C}^{EP}$ is the elastic–plastic matrix:

$$\mathbf{C}^{EP} = \mathbf{C}^{E} - \frac{9G^2}{(H' + 3G)\sigma_{eq}^2} \begin{bmatrix} \sigma'_{xx}\sigma'_{xx} & \sigma'_{yy}\sigma'_{xx} & \sigma'_{zz}\sigma'_{xx} & 2\sigma'_{xy}\sigma'_{xx} & 2\sigma'_{yz}\sigma'_{xx} & 2\sigma'_{zx}\sigma'_{xx} \\ \sigma'_{xx}\sigma'_{yy} & \sigma'_{yy}\sigma'_{yy} & \sigma'_{zz}\sigma'_{yy} & 2\sigma'_{xy}\sigma'_{yy} & 2\sigma'_{yz}\sigma'_{yy} & 2\sigma'_{zx}\sigma'_{yy} \\ \sigma'_{xx}\sigma'_{zz} & \sigma'_{yy}\sigma'_{zz} & \sigma'_{zz}\sigma'_{zz} & 2\sigma'_{xy}\sigma'_{zz} & 2\sigma'_{yz}\sigma'_{zz} & 2\sigma'_{zx}\sigma'_{zz} \\ \sigma'_{xx}\sigma'_{xy} & \sigma'_{yy}\sigma'_{xy} & \sigma'_{zz}\sigma'_{xy} & 2\sigma'_{xy}\sigma'_{xy} & 2\sigma'_{yz}\sigma'_{xy} & 2\sigma'_{zx}\sigma'_{xy} \\ \sigma'_{xx}\sigma'_{yz} & \sigma'_{yy}\sigma'_{yz} & \sigma'_{zz}\sigma'_{yz} & 2\sigma'_{xy}\sigma'_{yz} & 2\sigma'_{yz}\sigma'_{yz} & 2\sigma'_{zx}\sigma'_{yz} \\ \sigma'_{xx}\sigma'_{zx} & \sigma'_{yy}\sigma'_{zx} & \sigma'_{zz}\sigma'_{zx} & 2\sigma'_{xy}\sigma'_{zx} & 2\sigma'_{yz}\sigma'_{zx} & 2\sigma'_{zx}\sigma'_{zx} \end{bmatrix}. \tag{2.62}$$

In the former equation, $\mathbf{C}^{\mathrm{E}}$ represents the elasticity matrix (Eq. 2.13b), G is the shear modulus; H' is the slope of the material hardening function; and σ_{eq} is the equivalent stress, determined from the hardening function (Eq. 2.43a, b).

The Lagrangian finite element formulation is based on the equation:

$$\mathbf{K}^{(e)}\mathrm{d}\mathbf{U}^{(e)} = \mathrm{d}\mathbf{F}^{(e)}; \qquad (2.63)$$

where $\mathrm{d}\mathbf{U}^{(e)} = [\mathrm{d}u_{x1}, \mathrm{d}u_{y1}, \mathrm{d}u_{z1}, \ldots, \mathrm{d}u_{xN}, \mathrm{d}u_{yN}, \mathrm{d}u_{zN}]^{\mathrm{T}}$ is the vector of incremental nodal displacements; and $\mathrm{d}\mathbf{F}^{(e)} = [\mathrm{d}f_{x1}, \mathrm{d}f_{y1}, \mathrm{d}f_{z1}, \ldots, \mathrm{d}f_{xN}, \mathrm{d}f_{yN}, \mathrm{d}f_{zN}]^{\mathrm{T}}$ is the vector of incremental nodal forces. $\mathbf{K}^{(e)}$ is the element stiffness matrix, given by (2.61).

As the stiffness matrix involves the components of the deviatoric part of the stress tensor, equations (2.60) and (2.63) are nonlinear and cannot be solved straightforwardly but numerically integrated. This integration is usually performed by a generalized midpoint rule (Wriggers 2008):

$$u_{n+1} = u_n + \Delta t f(u_{n+\theta}); \qquad (2.64a)$$

$$u_{n+\theta} = (1 - \theta)u_n + \theta u_{n+1}, \quad 0 \le \theta \le 1. \qquad (2.64b)$$

For $\theta = 0$, this equations lead to the explicit Euler scheme; for $\theta = 1$, they lead to the implicit Euler scheme; finally, for $\theta = \frac{1}{2}$, they lead to the midpoint rule.

There is, also, a mixed approach, known as arbitrary Lagrangian–Eulerian (ALE) formulation, which has been widely used in recent years especially in modeling of cutting processes. It combines the features of pure Lagrangian analysis in which the mesh follows the material and Eulerian analysis where the mesh is fixed, when is needed as part of the adaptive remeshing.

In an ALE approach the process starts with an initial formed chip geometry, which is iteratively modified as the analysis proceeds until converging to the final shape of the chip (Vaziri et al. 2011).

Other approach in the ALE formulation allows the simulation of the chi formation, starting from zero, through a transient analysis. This method requires kinematic penalty contact conditions between the tool and the workpiece (Arrazola and Özel 2010).

A more detailed explanation on the ALE formulation, including the used algorithms and equations can be found in (Pantalé et al. 2004 and Olovsson et al. 1999).

2.4 Thermal Analysis

Thermal phenomena must be taken into account in FEM-based modeling of manufacturing processes, especially in machining and hot forming. Not only in the chip, but also in the tool, temperature, T, behavior is governed by the so-called Fourier's law (Li et al. 2002):

$$\kappa \nabla^2 T - \rho c \dot{T} + \dot{Q} = 0; \tag{2.65}$$

where ∇^2 is the Laplace's operator; κ and c are the thermal conductivity and specific heat of the material; and $\dot{Q}$ is the rate of generated heat.

There are two main sources of heat generation in processes involving metal working: plastic deformation and friction. The heat generated by plastic deformation can be computed by:

$$\dot{Q}_P = \eta_P \boldsymbol{\sigma} : \dot{\boldsymbol{\varepsilon}}^P; \tag{2.66}$$

being η_P the fraction of plastic work transformed into heat (usually $\eta_P \approx 0.9$); $\boldsymbol{\sigma}$ the Cauchy stress tensor and $\dot{\boldsymbol{\varepsilon}}^P$ the plastic strain tensor.

The heat generated by friction is given by:

$$\dot{Q}_f = \eta_f \tau_f v_s; \tag{2.67}$$

where η_f is the fraction of friction work transformed into heat (for machining applications $\eta_f \approx 1$), τ_f is the friction shear stress and v_s is the sliding velocity.

The heat flux, q, to the environment, from the free surfaces of the tool and part can be computed by the Newton's law of cooling (Grzesik 2006):

$$q = h(T_W - T_0); \tag{2.68}$$

where h is the convection heat transfer, T_W is the wall temperature and T_0 is the room temperature.

For computational efficiency, temperature discretization and computation are carried out together with the solution of the plasticity problem, in the so-called coupled formulation.

2.5 Friction Models

Friction is very important factor in most of the machining processes. Friction between the tool and chip, in cutting processes, has a strong influence on the forces and temperature and, consequently, on the operation economy. In forming processes, friction also plays a crucial role.

The simplest friction law is the so-called Coulomb's law, which considers a constant friction factor, μ, relating the friction shear stress, τ_f, and the normal stress acting on the surface, σ_n:

$$\tau_f = \mu \sigma_n. \tag{2.69}$$

Although in concordance with empirical data, this expression fails matching the friction behavior ah high pressures and sliding velocities, such as those taking place in cutting processes. Therefore, other more complex models have been proposed.

Coulomb's law sometimes contains a term, b, representing the cohesion sliding resistance (i.e., sliding resistance with zero normal pressure) (Pramanik et al. 2007):

Fig. 2.11 Stress distribution
at the tool rake face

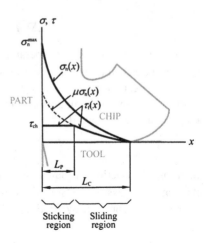

$$\tau_f = \mu\sigma_n + b; \tag{2.70}$$

although this parameter is usually neglected in machining simulations.

A modified Coulomb's law, proposed by Zorev (Vaziri et al. 2011) considers two different regions from the friction point of view. In the sliding region, where elastic contact exists, the value of the friction shear stress is proportional to the normal stress. In the sticking region, where plastic friction takes place, it is constant and equal to the average shear flow stress of the chip material on the chip-tool interface, τ_{ch}:

$$\tau_f = \begin{cases} \mu\sigma_n & : \mu\sigma_n < \tau_{ch} \ (0 \le x \le L_P) & \text{(sliding region)} \\ \tau_{ch} & : \mu\sigma_n \ge \tau_{ch} \ (L_P < x \le L_C) & \text{(sticking region)} \end{cases}. \tag{2.71}$$

The average shear flow stress in the sticking region can be considered as equal to the material yield shear stress, τ_Y (Zhang et al. 2011):

$$\tau_{ch} = \tau_Y = \frac{\sigma_Y}{\sqrt{3}}. \tag{2.72}$$

The distribution of the normal stress through the rake face (Fig. 2.11) can be approximately modeled by the following empirical relationship (Mohammadpour et al. 2010):

$$\sigma_n = \sigma_n^{max}\left[1 - \left(\frac{x}{L_C}\right)^a\right]; \tag{2.73}$$

where L_C is the chip-tool contact length, σ_n^{max} is the maximum normal stress and a is an empirical constant.

The Usui and Shirakashi's model (Filice et al. 2007) relates the friction shear stress with the normal stress and the flow shear stress by using a more complex expression:

$$\tau_f = \tau_{ch}\left[1 - \exp\left(-\frac{\mu\sigma_n}{\tau_{ch}}\right)\right]. \tag{2.74a}$$

which has been modified by Childs and coworkers by adding a proportionality term, m $(0 < m < 1)$ (Filice et al. 2007):

$$\tau_f = m\tau_{ch}\left[1 - \exp\left(-\frac{\mu\sigma_n}{\tau_{ch}}\right)\right]. \tag{2.74b}$$

Other interesting approach is the cohesive model (Mamalis et al. 2002), which constitutes an idealized model of the influence of the elastic-plastic deformation of the asperities at the metal surface at microscopic scale (cold weld phenomenon). This model establishes that:

$$\tau_f = -m\frac{2\sigma_{eq}}{3\sqrt{3}}\tan^{-1}\left(\frac{v_s}{c}\right); \tag{2.75}$$

where m is the shear friction factor; σ_{eq} is the equivalent stress, v_s is the sliding velocity and c is a constant representing the value of sliding velocity at which sliding occurs.

Sometimes, it results more convenient to obtain empirical expressions relation the friction factor with other parameters. For example, Rech et al. (2009) divide the apparent friction factor, μ_{app}, into a component due to the plastic deformation, μ_{plast}, and another one due to adhesive phenomena, μ_{adh}:

$$\mu_{app} = \mu_{plast} + \mu_{adh}; \tag{2.76}$$

having the adhesive component a linear dependency with the average local sliding velocity for a combination of annealed AISI 1045 steel and TiN-coated carbide.

Also for AISI 1045 steel and uncoated carbide, Brocail et al. (2010) have obtained, through FEM simulation, the exponential model:

$$\mu = 0.919\sigma_n^{-0.251}v_s^{-0463}T_{int}^{0.480}; \tag{2.77}$$

relating the friction factor, μ, with the normal stress, σ_n, the sliding velocity, v_s, and the temperature at the interface, T_{int}.

In spite of the success of these models in some cases, more experimental and theoretical research is required in this field, in other to obtain actually reliable and flexible friction models.

2.6 Fracture

In Lagrangian and ALE formulation there is a need of use some fracture criterion which allows evaluating the state of damage of the elements and carries out the separation of the chip in cutting processes and the breakage of the workpiece in forming.

Nearly all the strain–stress based fracture criteria came from the Freudenthal's criterion, which considers that fracture takes place when the plastic work per unit volume reaches some critical value, C_1 (Gouveia et al. 2000):

$$\int_0^{\varepsilon_{eq}^F} \sigma_{eq} d\varepsilon_{eq} = C_1; \tag{2.78}$$

where σ_{eq} and ε_{eq} are the equivalent stress and strain, respectively, and ε_{eq}^F is the value of equivalent strain at which the fracture take place.

The normalized Cockcroft-Latham criterion uses the relation between the largest principal stress, σ_1, and the equivalent stress, σ_{eq} (Umbrello 2008):

$$\int_0^{\varepsilon_{eq}^F} \frac{\sigma_1}{\sigma_{eq}} d\varepsilon_{eq} = C_2; \tag{2.79}$$

while the Brozzo's equation combines the effects of the principal stress, σ_1, and the hydrostatic stress, p (Gouveia et al. 2000):

$$\int_0^{\varepsilon_{eq}^F} \frac{2\sigma_1}{3(\sigma_1 - p)} d\varepsilon_{eq} = C_3. \tag{2.80}$$

Another approach (Rosa et al. 2007), which is based on the specific distortion energy, states that the ductile damage takes place when:

$$\int_0^{\gamma^F} \tau d\gamma = C_4; \tag{2.81}$$

where τ and γ are the shear stress and the distortion of the element, and γ^F is the level of material distortion at the onset of cracking.

The stress index parameter is also used as a fracture criterion (Shet and Deng 2003). It is defined as:

$$f = \sqrt{\left(\frac{\sigma_n}{\sigma_F}\right)^2 + \left(\frac{\tau}{\tau_F}\right)^2}; \tag{2.82a}$$

where σ_F and τ_F are the failure stresses of the material under pure tensile and shear loading conditions. The fracture starts when the stress index parameter reaches the value of one:

$$f \geq 1. \tag{2.83b}$$

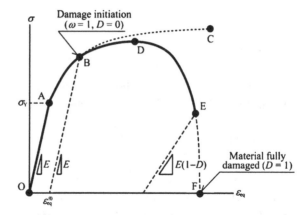

Fig. 2.12 Typical stress–strain material response with damage

A more elaborated damage approach is the Johnson–Cook's shear failure model (Zhang et al. 2011) with damage starting at point B (see Fig. 2.12), where the cumulative scalar parameter:

$$\omega = \sum_{j=1}^{n} \left(\frac{\Delta \varepsilon_{eq}^{P}}{\varepsilon_{eq}^{f0}} \right)_{j}; \tag{2.84}$$

exceeds 1. In Eq. (2.84) $\Delta \varepsilon_{eq}^{P}$ is the increment of equivalent plastic strain in the computation step j, and:

$$\varepsilon_{eq}^{f0} = \left[d_1 + d_2 \exp\left(d_3 \frac{p}{\sigma_{eq}} \right) \right] \left[1 + d_4 \ln\left(\frac{\dot{\varepsilon}_{eq}}{\dot{\varepsilon}_{eq}^{0}} \right) \right] \left[1 - d_5 \left(\frac{T - T_0}{T_M - T_0} \right) \right]; \tag{2.84}$$

is the equivalent strain at the damage initiation, p is the hydrostatic pressure, σ_{eq} is the equivalent stress, $\dot{\varepsilon}_{eq}$ is the equivalent strain rate, $\dot{\varepsilon}_{eq}^{0}$ is the reference strain rate, T is the material temperature, T_0 is the reference temperature, T_M is the material melting temperature, and d_i ($i = 1, \ldots, 5$) are material constants.

The damage evolution (line B–F) can be characterized by the scalar stiffness degradation, D, which is equal to zero at the damage initiation (point B) and is equal to one at the theoretical final fracture (point F). This parameter can be computed by the expression:

$$D = \frac{\int_{0}^{u_{eq}^{P}} \sigma_{eq} du_{eq}^{P}}{G_f}; \tag{2.85a}$$

for the exponential damage zone, and by the expression:

$$D = 1 - \exp\left(-\int\limits_0^{u_{eq}^P} \frac{\sigma_{eq}}{G_f} \, du_{eq}^P\right);$$ (2.85b)

for the linear damage zone. In both equations, u_{eq}^P is the equivalent plastic displacement and G_f is the Hillerborg's fracture energy:

$$G_f = \int\limits_0^{u_{eq}^P} \sigma_{eq} du_{eq}^P.$$ (2.86)

The effectiveness of all these models varies notably under different conditions and depends on empirical constant whose experimental determination imposes serious constraints to their use. However, successful application of these approaches has been done in the area of manufacturing FEM-based processes modeling.

References

P.J. Arrazola, T. Özel, Investigations on the effects of friction modeling in finite element simulation of machining. Int. J. Mech. Sci. **52**, 31–42 (2010). doi:10.1016/j.ijmecsci.2009.10.001

K.-J. Bathe, *Finite Element Procedures* (Prentice Hall, Upper Saddle River, 1996)

J. Brocail, M. Watremez, L. Dubar, Identification of a friction model for modelling of orthogonal cutting. Int. J. Mach. Tools Manuf. **50**, 807–814 (2010). doi:10.1016/j.ijmachtools.2010.05.003

J. Chakrabarty, *Theory of Plasticity*, 3rd edn. (Elsevier Butterworth-Heinemann, Oxford, 2006)

P.M. Dixit, U.S. Dixit, *Modeling of Metal Forming and Machining Processes by Finite Element and Soft Computing Methods* (Springer, London, 2008)

U.S. Dixit, S.N. Joshi, J.P. Davim, Incorporation of material behavior in modeling of metal forming and machining processes: A review. Mater. Des. **32**, 3655–3670 (2011). doi:10.1016/j.matdes.2011.03.049

L. Filice, F. Micari, S. Rizzuti, D. Umbrello, A critical analysis on the friction modelling in orthogonal machining. Int. J. Mach. Tools Manuf. **47**, 709–714 (2007). doi:10.1016/j.ijmachtools.2006.05.007

B.P.P.A. Gouveia, J.M.C. Rodrigues, P.A.F. Martins, Ductile fracture in metalworking: experimental and theoretical research. J. Mater. Process. Tech. **101**, 52–63 (2000). doi:10.1016/S0924-0136(99)00449-5

W. Grzesik, Determination of temperature distribution in the cutting zone using hybrid analytical-FEM technique. Int. J. Mach. Tools Manuf. **46**, 651–658 (2006). doi:10.1016/j.ijmachtools.2005.07.009

W. Han, B.D. Reddy, *Plasticity: Mathematical Theory and Numerical Analysis* (Springer, New York, 1999)

S.P.F.C. Jasper, J.H. Dautzenberg, Material behaviour in conditions similar to metal cutting: flow stress in the primary shear zone. J. Mater. Process. Tech. **122**, 322–330 (2002). doi:10.1016/S0924-0136(01)01228-6

D.I. Lalwani, N.K. Mehta, P.K. Jain, Extension of Oxley's predictive machining theory for Johnson and Cook flow stress model. J. Mater. Process. Tech. **209**, 5305–5312 (2009). doi:10.1016/j.jmatprotec.2009.03.020

K. Li, X.-L. Gao, J.W. Sutherl, Finite element simulation of the orthogonal metal cutting process for qualitative understanding of the effects of crater wear on the chip formation process. J. Mater. Process. Tech. **127**, 309–324 (2002). doi:10.1016/S0924-0136(02)00281-9

G.R. Liu, S.S. Quek, *The Finite Element Method: A Practical Course* (Butteeworth-Heinemann, Burlington, 2003)

A.G. Mamalis, A.S. Branis, D.E. Manolakos, Modelling of precision hard cutting using implicit finite element methods. J. Mater. Process. Tech. **123**, 464–475 (2002). doi:10.1016/S0924-0136(02)00133-4

M. Mohammadpour, M.R. Razfar, R.J. Saffar, Numerical investigating the effect of machining parameters on residual stresses in orthogonal cutting. Simul. Model Pract. Theory **18**, 378–389 (2010). doi:10.1016/j.simpat.2009.12.004

L. Olovsson, L. Nilsson, K. Simonsson, An ALE formulation for the solution of two-dimensional metal cutting problems. Comput. Struct. **72**, 497–507 (1999)

T. Özel, E. Zeren, Determination of work material flow stress and friction for FEA of machining using orthogonal cutting tests. J. Mate.r Process. Tech. **153–154**, 1019–1025 (2004). doi:10.1016/j.jmatprotec.2004.04.162

O. Pantalé, J.-L. Bacaria, O. Dalverny, R. Rakotomalala, S. Caperaa, 2D and 3D numerical models of metal cutting with damage effects. Comput. Method Appl. M. **193**, 4383–4399 (2004). doi:10.1016/j.cma.2003.12.062

A. Pramanik, L.C. Zhang, J.A. Arsecularatne, An FEM investigation into the behavior of metal matrix composites: Tool–particle interaction during orthogonal cutting. Int. J. Mach. Tools Manuf. **47**, 1497–1506 (2007). doi:10.1016/j.ijmachtools.2006.12.004

J. Rech, C. Claudin, E. D'Eramo, Identification of a friction model: Application to the context of dry cutting of an AISI 1045 annealed steel with a TiN-coated carbide tool. Tribol. Int. **42**, 738–744 (2009). doi:10.1016/j.triboint.2008.10.007

P.A.R. Rosa, O. Kolednik, P.A.F. Martins, A.G. Atkins, The transient beginning to machining and the transition to steady-state cutting. Int. J. Mach. Tools Manuf. **47**, 1904–1915 (2007). doi:10.1016/j.ijmachtools.2007.03.005

A. Shabana, *Computational Continuum Mechanics* (Cambridge University Press, Cambridge, 2008)

C. Shet, X. Deng, Residual stresses and strains in orthogonal metal cutting. Int. J. Mach. Tools Manuf. **43**, 573–587 (2003). doi:10.1016/S0890-6955(03)00018-X

D. Umbrello, Finite element simulation of conventional and high speed machining of Ti6Al4V alloy. J. Mater. Process. Tech. **196**, 79–87 (2008). doi:10.1016/j.jmatprotec.2007.05.007

M.R. Vaziri, M. Salimi, M. Mashayekhi, Evaluation of chip formation simulation models for material separation in the presence of damage models. Simul. Model Pract. Theory **19**, 718–733 (2011). doi:10.1016/j.simpat.2010.09.006

P. Wriggers, *Nonlinear Finite Element Methods* (Springer, Berlin, 2008)

Y.C. Zhang, T. Mabrouki, D. Nelias, Y.D. Gong, Chip formation in orthogonal cutting considering interface limiting shear stress and damage evolution based on fracture energy approach. Finite Elem. Anal. Des. **47**, 850–863 (2011). doi:10.1016/j.finel.2011.02.016

O.C. Zienkiewicz, R.L. Taylor, *The Finite Element Method: The Basis* (Butterworth-Heinemann, Oxford, 2000)

Chapter 3
Artificial Intelligence Tools

Abstract This chapter summarizes the main concepts on artificial intelligence, remarking those tools which are commonly applied to the modeling and optimization of manufacturing processes. Special emphasis has been done on soft computing techniques, because of the wide use that these ones have in this field. Each of the main soft computing techniques (artificial neural networks, fuzzy logic and stochastic optimization) is explained and, examples of applications are given.

3.1 Preliminary Concepts

Artificial intelligence (AI) is a branch of computer science aiming to study and design intelligent agents, which are systems that perceive their environment and take actions that maximize their chance of success. Created in the middle of the twentith century, with the purpose of simulate the human reasoning capabilities, AI has followed a discontinuous development path, with moments of achievements, such as the commercial introduction of expert system in the early 1980s or the massive application of specific techniques in industry and services, in the 1990s and early twenty first century, and periods of stagnation, as in the 1970s, after an enthusiastic beginning.

Currently, although still failing in its initial goal of create intelligent machines, AI has reached a remarkable success in dealing with specific problems with very high complexity, for which conventional methods do not provide low cost, complete and analytical solutions. These AI-based approaches are commonly known as soft computing, in order to be differentiated from the classic AI hard computing techniques. Soft computing is more based in biological systems than in formal logic, and it uses inductive reasoning more extensively than hard computing schemes.

R. Quiza et al., *Hybrid Modeling and Optimization of Manufacturing*,
SpringerBriefs in Computational Mechanics,
DOI: 10.1007/978-3-642-28085-6_3, © The Author(s) 2012

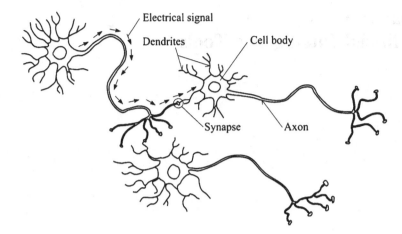

Fig. 3.1 Interconnected neurons

Soft computing includes, but is not limited to, neural networks, fuzzy logic, evolutionary computation, swarm intelligence, probability-based systems and chaos theory. In the following sections, these techniques will be briefly explained.

3.2 Artificial Neural Networks

3.2.1 Biological Foundations and Neuron Model

Artificial neural networks are connectionist structures originally created for simulating the pattern recognition capabilities of the human brain. However, at the present they have found successful applications in several branches of the science and technique.

Human brain is composed by a very large number of cells, called neurons, which are highly interconnected. As an average, there are 10^{11} neurons in a human brain, each of them with 10^4 connections with other ones (Hagan et al. 2002). In spite of its complex biological structure, for modeling purposed, a neuron can be considered as formed by the cell body, the dendrites, which are tree-like fiber elements that carry the electrical signals into the cell body, and the axon, which is a single long fiber that transports the electrical signals from the cell body out to the other neurons (see Fig. 3.1). The contact between the axon of one neuron and a dendrite of another one is called a synapse. The strength of these synapses determines how the signal is transported from one neuron to other one and it is given by complex biochemical processes.

The most accepted mathematical model for a neuron was proposed by McCulloch and Pitts (Hu and Hwang 2002). In this model, the neuron is considered having several inputs, $x_1, x_2, \ldots, x_N$, and one output, y (see Fig. 3.2).

Fig. 3.2 McCulloch and
Pitts' neuron model

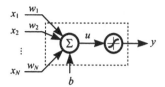

The inputs of the neuron are first aggregated for obtaining a combined
input, u. The most used way for doing that is through a linear combination in the form:

$$u = b + \sum_{i=1}^{N} w_i x_i; \qquad (3.1a)$$

where the weights, $w_1, w_2, ..., w_N$, represent the strength of the synapses and b is
the so-called bias or threshold, which corresponds to the predisposition of the
neuron for being activated.

It becomes very convenient consider the threshold, b, as another weight, w_0,
carrying a unit input $x_0 = 1$ into the neuron, so that the aggregated input takes the
form:

$$u = \sum_{i=0}^{N} w_i x_i. \qquad (3.1b)$$

After that, the output is computed from the combined input by using some
activation function:

$$y = f(u). \qquad (3.2)$$

Several activation functions have been proposed for different kinds of neural
networks. Sigmoid (Fig. 3.3a) and hyperbolic tangent (Fig. 3.3b) present the
advantage of having derivatives than can be computed directly from the value of
the functions. Arctangent (Fig. 3.3c) has a similar behavior than hyperbolic
tangent but with a slower increment rate. It is not frequently used. Linear function
(Fig. 3.3d) has a constant derivative but is not bounded. Step or threshold function
(Fig. 3.3e) has been widely used in classification tasks, but is not very convenient
for functional approximation. Finally, Gaussian radial basis function (Fig. 3.3f) is
used in radial basis function networks.

3.2.2 Network Topology and Learning

A neural network consists in a certain number of neurons interconnected in some
mode or topology. These topologies can be conveniently represented by oriented
graphs, where the nodes represent neurons and the directed arcs represent synapses Fig. 3.4. Neurons are usually arranged in $M + 1$ layers, each of them having

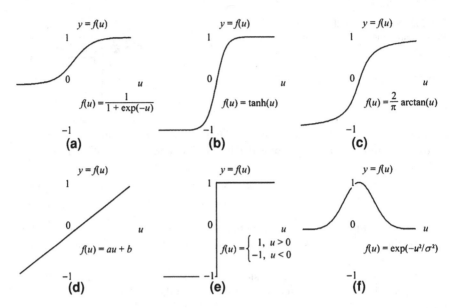

Fig. 3.3 Activation functions

certain number of neurons, $K^{(0)}$, $K^{(1)}$, ..., $K^{(M)}$. The first layer (number cero) is known as input layer; it is established only for organizational purposes and hardware implementation and no computations are performed in it. As evident, the number of neuron in this layer, $K^{(0)}$, corresponds to the number of network inputs, $[X_1, X_2, ..., X_{K(0)}]^T$. The output layer give the $K^{(M)}$ outcomes of the network, $[Y_1, Y_2, ..., Y_{K(M)}]^T$. Finally, layers from 1 to $M-1$ receive the denomination of hidden layers because their inputs and outputs are not shown to the environment.

If there is not any closed loop in the graph representing the network, i.e. if there is not any connection from the output of one neuron to the input to a neuron belonging to a previous layer, the network is called feed-forward, non-recurrent or acyclic (see Fig. 3.5a). Examples of this kind of topology are the multilayer perceptron (MLP), the radial basis function network (RBFN) and the self-organizing maps (SOM).

On the contrary, if there is a closed loop in the graph, the network is called recurrent or cyclic (Fig. 3.5b). These networks are often applied in modeling nonlinear dynamic systems. Examples of this topology are the Elman network and the Hopfield network.

One of the fundamental problems in a neural network is how to set the weights and biases in order to obtain the desired behavior of the network. This is done from some previously known collection of inputs and, sometimes, outputs, which form the so-called training set and the process is named training or learning.

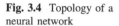

Fig. 3.4 Topology of a
neural network

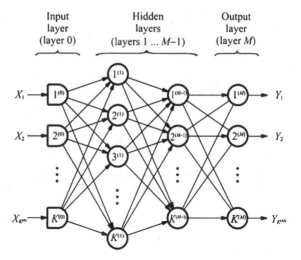

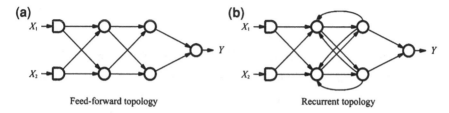

Fig. 3.5 Feed-forward versus. recurrent networks

There are three main kinds of learning procedures (Hagan et al. 2002):

- Supervised learning, where not only the input but also the output for every sample in the training set is previously known.
- Reinforcement learning, which, instead of actual values of the correct output for every element of the training set is provided with some measure of the network performance.
- Unsupervised learning, where there are not available output, so the weights and biases are modified in response to the inputs only.

3.2.3 Multilayer Perceptron

Multilayer perceptron (MLP) is not only the most studied paradigm into the neural networks, but also by far the most widely used. It is based in the perceptron introduced by Rosenblatt in the 1950s. The original perceptron was a neuron with a linear aggregator for the inputs and a threshold activation function. While it showed being effective in mapping separable sets, it is not capable of mapping non-linearly separable sets, such as in a XOR function (Tajine and Elizondo 2002).

The MLP is a feed-forward layered network of perceptrons. Instead of threshold, MLP often used sigmoid or arctangent activation functions, although linear function is also used especially in the output layer. It has been proved that a MLP with as few as two hidden layers is capable of approximating an arbitrary complex mapping within a finite support (Hu and Hwang 2002).

The training process of a MLP is based on the error back-propagation algorithm Rajasekaran and Vijayalakshmi Raj (2003). It is a supervised learning rule, which is carried out in two iterative steps. First, the weights and biases are randomly initialized. Then, the training samples are presented one at a time. Thence, for each sth sample, the output of each neuron is computed from its inputs:

$$y_j^{(k)}(s) = f[u_j^{(k)}(s)]; \tag{3.3a}$$

being:

$$u_j^{(k)}(s) = \sum_{i=0}^{K^{(k-1)}} w_{i,j}^{(k)} x_i^{(k)}(s); \tag{3.3b}$$

where $f(\bullet)$ is the activation function; $y_j^{(k)}(s)$, the output of the jth neuron in the kth layer for the sth sample; $b_j^{(k)}$, the bias of this neuron; $w_{i,j}^{(k)}$, the weight connecting the ith neuron in the $(k-1)$th layer with the jth neuron in the kth layer; and $x_i^{(k)}(s)$, the input coming from the ith neuron in the $(k-1)$-h layer for the sth sample.

As the output of a neuron serves as input of the following ones:

$$y_i^{(k-1)}(s) = x_i^{(k)}(s); \tag{3.4}$$

this calculation is done from the first layer to the last one, i.e., in the forward direction Fig. 3.6.

After that, in the second step, the error, $\delta_j^{(k)}(s)$, is computed for each neuron, beginning in the last layer and going to the first one, that is to say, in the backward direction. For the neurons in the last layer ($k = M$) the error is computed directly by the difference between the computed output, $y_j^{(M)}(s)$, and the desired one, Y_j:

$$\delta_j^{(M)}(s) = Y_j - y_j^{(M)}(s). \tag{3.5}$$

In the hidden layers ($k = 1 \dots M-1$), the error is obtained by applying the steepest descend gradient method, by using:

$$\delta_j^{(k)}(s) = f'[u_i^{(k)}(s)] \sum_{i=1}^{K^{(k+1)}} \delta_i^{(k+1)}(s) w_{j,i}^{(k+1)}; \tag{3.6}$$

where $f'(\bullet)$ is the first derivative of the activation function.

Fig. 3.6 Forward step in a neuron

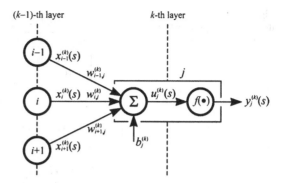

Fig. 3.7 Backward step in a neuron

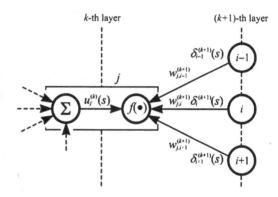

With the errors of each neuron, computed for each learning sample, at the epoch t, the weights, for the epoch $t + 1$, are updated through the following expression:

$$w_{i,j}^{(k)}(t+1) = w_{i,j}^{(k)}(t) + \eta \sum_{s=1}^{S} \left[\delta_j^{(k)}(s) y_i^{(k-1)}(s) \right] + \ldots$$

$$\ldots + \mu \left[w_{i,j}^{(k)}(t) - w_{i,j}^{(k)}(t-1) \right] + \varepsilon_{i,j}^{(k)}(t); \qquad (3.7)$$

where η (usually in the interval 0 ... 0.3) is the learning rate; μ (usually in the interval 0.6 ... 0.9) is the momentum constant and $\varepsilon_{i,j}^{(k)}(t)$ is a small random noise term which is useful for avoiding the convergence of the learning process to a local minimum Fig. 3.7.

This process is repeated until achieving some preset value of sum of squared errors. In Fig. 3.8, a simplified block diagram of the error back-propagation algorithm is presented. Some improvements of this algorithm, such as the adaptive learning rate (Behera et al. 2006) and the inclusion of momentum have been developed (Zhang 2009).

Fig. 3.8 Block diagram of
the error back-propagation
algorithm

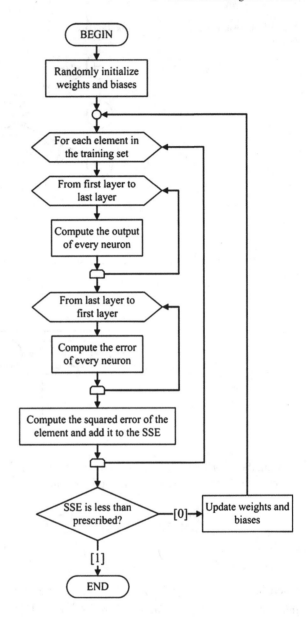

3.2.4 Radial Basis Function Networks

Radial basis function networks (RBFNs) are another important kind of feed forward neural networks which have been widely used in modeling and optimization of manufacturing processes. They are based on a special kind of activation function called radial basis function (RBF) because their value changes with the

Fig. 3.9 RBFN structure

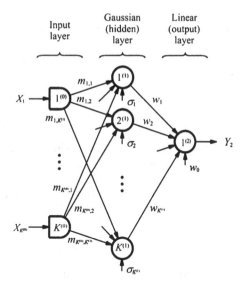

distance to some fixed point (Buhmann 2004). The most commonly used RBF is the Gaussian:

$$f(\mathbf{x}) = \exp\left(\frac{-||\mathbf{x} - \mathbf{m}||^2}{2\sigma^2}\right); \tag{3.8}$$

where $\mathbf{x} = [x_1, x_2, \ldots, x_K]^T$ is the vector of the inputs; $\mathbf{m} = [m_1, m_2, \ldots, m_K]^T$ is the vector of the coordinates of the fixed point also known as center; and σ is a parameter called spread.

A RBFN is composed by two layers (see Fig. 3.9). The first one uses Gaussian activation functions, where the coordinates of the centers are the weights of the neurons and the spreads, the biases. Therefore, the output of the jth neuron of the first layer can be computed by using (Roussos and Baxter 2005):

$$y_j^{(1)} = \exp\left(-\frac{1}{2\sigma_j^2}\sum_{i=1}^{K^{(0)}}(x_i - m_{i,j})^2\right). \tag{3.9}$$

On the contrary, the second (output) layer has linear activation functions, and its output can be obtained by:

$$y^{(2)} = \sum_{i=0}^{K^{(1)}} w_i x_i^{(2)}. \tag{3.10}$$

It is possible, but not common, to have more than one output in the RBFN. In that case, the equation (3.10) is applied for every output neuron.

The learning process of the RBFN, although supervised, is quite different from back-propagation. In a first approach, the center points are selected equal to the samples in the training set and the value of spread is also previously established. Then, only the weights w_i are the only part of the network to adapt during the learning process.

As the output layer is linear, the minimization of the sum squared error:

$$\text{min } SSE = \sum_{s=1}^{S} \left[Y(s) - \sum_{i=0}^{K^{(1)}} w_i x_i^{(2)}(s) \right]^2 ; \qquad (3.11)$$

leads to a system of linear equations, which can be directly solved. If the design matrix, $\mathbf{H}$, is defined as:

$$\mathbf{H} = \begin{bmatrix} f_1[\mathbf{x}(1)] & f_2[\mathbf{x}(1)] & \cdots & f_3[\mathbf{x}(1)] \\ f_1[\mathbf{x}(2)] & f_2[\mathbf{x}(2)] & \cdots & f_3[\mathbf{x}(2)] \\ \vdots & \vdots & & \vdots \\ f_1[\mathbf{x}(3)] & f_2[\mathbf{x}(3)] & \cdots & f_3[\mathbf{x}(3)] \end{bmatrix} ; \qquad (3.12)$$

where $f_i[\mathbf{x}(j)]$ is the radial basis activation function for the ith hidden neuron evaluated by the jth sample of the learning set, the values of the weights, $\mathbf{w} = [w_1, w_2, ..., w_{K(2)}]^T$, can be computed by doing:

$$\mathbf{w} = (\mathbf{H}^T \mathbf{H})^{-1} \mathbf{H}^T \mathbf{Y}. \qquad (3.13)$$

The most interesting particularity of this kind of RBFNs is that they can predict exactly the values of the outputs of the training samples; therefore they are frequently referred as exact RBFN. Nevertheless, they often have a very poor generalization capability, i.e., they cannot predict accurately those values which are not contained in the training set.[1] Furthermore, when there are a large number of samples in the training set, the network requires too many hidden neurons, and consequently, losses computational efficiency.

One alternative for this situation is to use a number of centers less than the total amount of training samples. This give a BRFN which does not predict exactly the outputs for training samples, but it is smaller and has better generalization capabilities.

For doing that, the network is initialized with only one neuron. Then, neurons are added, one at a time, until the sum squared error of the network is lower than some prescribed value. In Fig. 3.10, the block diagram of this learning algorithm is shown.

Compared with MLPs, RBFNs can be trained faster and easier. However, they tend to have more neurons than a comparable MLP. The selection of one paradigm or the other one, depend on the size and characteristics of the problem.

---

[1] This problem, which is not exclusive of the RBFN, is explained more detailedly in Sect. 3.2.7.

Fig. 3.10 RBFN training

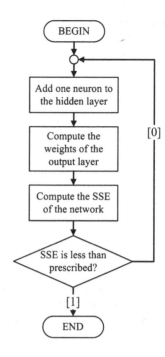

3.2.5 Hopfield Networks

Hopfield networks, proposed by John Hopfield, in 1982, are one of the most well known recurrent neural networks paradigms (Rojas 1996). They can take some input $\mathbf{x} = [x_1, x_2, \ldots, x_K]^T$ and approximate it to the most closed pattern from the training set $\mathbf{X} = [\mathbf{X}(1), \mathbf{X}(2), \ldots, \mathbf{X}(S)]$.

A Hopfield network (see Fig. 3.11) is composed by K neurons, all of them interconnected with the other ones (by synapses with weights w_{ij}), except with itself ($w_{ii} = 0$). In order to guaranty the stability of the network, the weights are symmetric, i. e., $w_{ij} = w_{ji}$. Additionally, each neuron has a bias, b_i (Kasabov 1998).

Neurons use a sign activation function:

$$y_i = \mathrm{sgn}\left(\sum_{j=1}^{K} w_{ij}x_i - b_i\right);$$

(3.14)

although sometimes other functions such as the step or the saturated linear are used instead. As the output of each neuron acts as input of the other ones, the computation process must be repeated until some equilibrium state is reached. This equilibrium corresponds to the minimum value of the energy:

$$E = -\frac{1}{2}\mathbf{x}^T\mathbf{W}\mathbf{x} + \mathbf{b}^T\mathbf{x};$$

(3.15)

Fig. 3.11 Structure of a
Hopfield network

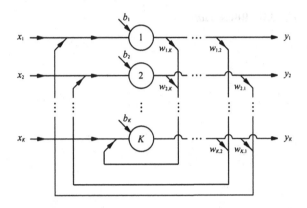

where $\mathbf{W} = \{w_{ij}\}$ is the symmetric matrix of weights and $\mathbf{b} = [b_1, b_2, ..., b_K]^T$ is
the biases vector.

For networks without biases (i. e., were the biases are just zero) the Hebbian
learning can be used. This method is applied for a training set of S selected of
K-dimensional stable states, $\mathbf{X}(s) = [X_1(s), X_2(s), ..., X_K(s)]^T, s = 1, 2, ..., S$. The
weights are computed by following the expression:

$$\mathbf{W} = \sum_{s=1}^{S} \mathbf{X}(s)\mathbf{X}(s)^T - S\mathbf{I}; \tag{3.16}$$

where $\mathbf{I}$ is the identity matrix.

In spite of its simplicity, Hebbian rule sometimes fails in finding the weight
matrix in Hopfield network, for which the S given vectors are stable states, even if
this matrix exits. This problem takes place especially when the vectors in the
training set lies near each other. In order to overcome this drawback, another
learning rule, which is a variation of the perceptron rule, is frequently used
(Rojas 1996).

In order to apply the perceptron learning rule, each pattern vector, $\mathbf{x}$, is transformed into K auxiliary vectors, $\mathbf{z}_1, \mathbf{z}_2, ..., \mathbf{z}_K$, of dimension $K + K(K-1)/2$:

$$\mathbf{z}_1 = [\underbrace{x_2, x_3, ..., x_K}_{K-1}, 0, ..., 0, \underbrace{1, 0, ..., 0}_{K}]^T$$

$$\mathbf{z}_2 = [\underbrace{x_1, 0, ..., 0}_{K-1}, \underbrace{x_3, ..., x_K}_{K-2}, 0, ..., 0, \underbrace{0, 1, 0, ..., 0}_{K}]^T$$

$$\vdots$$

$$\mathbf{z}_K = [\underbrace{0, ..., 0, x_1}_{K-1}, \underbrace{0, ..., 0, x_2}_{K-2}, 0, ..., 0, \underbrace{0, 0, ..., 0, 1}_{K}]^T \tag{3.17}$$

Fig. 3.12 Perceptron model equivalent to a Hopfield network

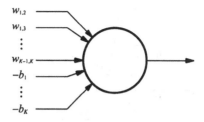

Then, an equivalent perceptron (see Fig. 3.12) is constructed with $K + K(K - 1)/2$ inputs, corresponding to the non-diagonal elements of the weight matrix and the biases, which form the vector:

$$\mathbf{v} = [\underbrace{w_{1,2}, w_{1,3}, \ldots, w_{1,K}}_{K-1}, \underbrace{w_{2,3}, w_{2,4}, \ldots, w_{2,K}}_{K-2}, \ldots, \underbrace{w_{K-1,K}}_{1}, \underbrace{-b_1, -b_2, \ldots, -b_K}_{K}]^{\mathrm{T}}.$$

(3.18)

If there are S training samples, the learning process is carried out with the $K \times S$ auxiliary vectors in order to obtain the components of $\mathbf{v}$ that perform a linear separation of the vectors $\mathbf{z}_1$, $\mathbf{z}_2$, ..., $\mathbf{z}_K$. After that, weights and biases for the Hopfield network are extracted from $\mathbf{v}$.

3.2.6 Adaptive Resonance Theory and Self-Organizing Maps

In classification tasks usually there is not information about the class where each sample is contained and, therefore, this is part of the task to be solved. For this kind of problems, unsupervised learning techniques have been designed.

One of the most popular neural network paradigms using unsupervised learning is based on adaptive resonance theory (ART), which was developed by Stephen Grossberg and Gail Carpenter in the middle seventies. The ART neural networks are ideal for conceptualization, clustering, and discovering existing types and number of classes in a database (Kasabov 1998). They were designed specifically to overcome the stability-plasticity dilemma (i.e., keeping the system responsive to new inputs while preserving the effects of pass inputs) (Harvey 1994).

ART-1, was the first ART-based approach, developed for dealing with binary vectors. The main clustering algorithm of the ART-1, independently on its implementation in any neural structure is shown in Fig. 3.13 (Heins and Tauirtz 1995). As can be seen, the first step is the initialization of the vigilance parameter ρ, which defines the class size. If this parameter is selected too close to zero, it produced large classes. As this parameter gets greater, classes become finer. In the extreme case, when it is equal to one, one class is defined for each input vector. The set of prototype vectors is also initialized as an empty set.

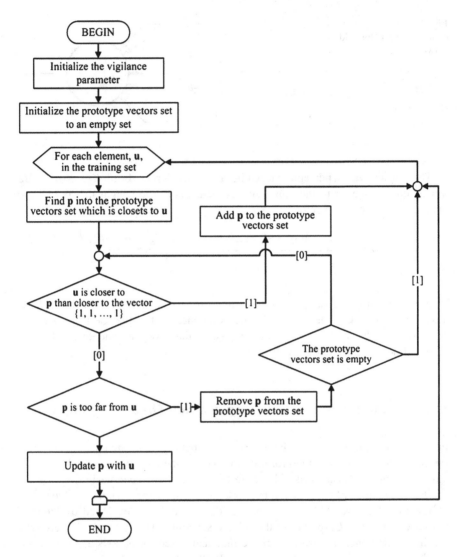

Fig. 3.13 Algorithm of an ART-1 based neural network

After the initialization, each sample **u** in the training set vectors is consecutively presented to the algorithm and a copy of the current prototype set vector is created. Then, it is found the prototype vector **p** closest to the current training vector **u**. As these vectors are binary, the distance, $d_{u/p}$, between them is defined as the magnitude (i.e., the number of ones that the vector contains) of the vector resulting from the bitwise AND operation (which can also be seen as a dot product, for binary vectors):

Fig. 3.14 Representation of an ART-1 network (Heins and Tauirtz 1995)

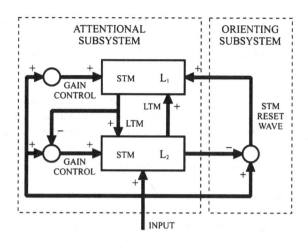

$$d_{\mathbf{u}/\mathbf{p}} = \frac{\|\mathbf{u} \cdot \mathbf{p}\|}{\beta + \|\mathbf{p}\|}. \tag{3.19}$$

In this comparison the parameter β acts as tie-breaker, favoring larger magnitude prototype vectors when multiple prototype vectors are subsets of the training vector. This compensates for the fact that prototype vectors can only move in one direction.

In the next step, it is verified if the current input vector, $\mathbf{u}$, is closer to the nearest prototype vector, $\mathbf{p}$, or to the vector formed by ones, $\{1, 1, ..., 1\}$. In this case, new prototype vector created equal to $\mathbf{u}$ and it is added to the prototype vector set, and the next input vector is presented.

If the before-mentioned condition is not fulfilled, then it is checked that the nearest prototype vector, $\mathbf{p}$, is not too far from $\mathbf{u}$. If it is, $\mathbf{p}$ is removed from the prototype vector set, and, if the set is not empty, a new nearest vector is searched; else, the new step in the loop is initiated by presenting the next input vector.

Finally, if $\mathbf{p}$ is not too far from $\mathbf{u}$, the prototype vector, $\mathbf{p}$, is updated through the application of the bitwise AND operation with $\mathbf{u}$, and the next loop step begins Fig. 3.14.

Though ART-1 is unsupervised it can sometimes be useful to add a limited amount of supervision by allowing the vigilance parameter to be changed externally.

An ART-1 network consists in two layers. In the first one, F_1, the input is characterized by extracting its representative features. Then, in the second layer, F_2, the category of this input is recognized based on its features.

Pattern of activation of F_1 and F_2 nodes are called sort-term memory (SMT) traces, because they exist only during a single presentation of an input vector. Layers F_1 and F_2 are fully connected through the weighted bottom-up and top-down pathways, which are known as long-term memory (LTM) traces and are equivalent to prototype vectors in the previously presented clustering algorithm.

When it is presented through the training process, the input pattern **u** generates the STM activity pattern **x** at the input layer F_1 nodes and also activates the gain control G_1 and inhibits the orienting subsystem A. The bottom-up signal, **s**, generated from **x** is transformed by the bottom-up pathway into de pattern **t**, which is presented to F_2.

As F_2 is a competitive layer, only the node which receives the larger total input is activated, creating the STM pattern **y**. Then, **y** generates the top-down signal pattern **w** which is transformed by the top-down pathway into the expectation pattern **v**. Pattern **y** also inhibits the gain control G_1. Consequently, only those F_1's nodes that represent bits in the intersection of the input pattern **u** and the expectation pattern **v** remain activated.

If **v** mismatches **u**, this causes a decrease in the total inhibition from F_1 to the gain control G_1. If the mismatch is severe enough, the gain control resets the active node at F_2. The vigilance parameter, ρ, determines how much mismatch will be allowed. The parallel search repeats until one of the following alternatives occurs: a F_2 node is chosen whose expectation **v** approximately matches input **u**; a previously uncommitted F_2 node is selected; or the entire capacity of the system is used and input u cannot be accommodated. Significant learning in response to an input pattern occurs when the cycle that it generates comes to an end (it is known as resonance state) (Heins and Tauirtz 1995).

Later on, some improvement of the ART networks have been developed, such as ART-2 which can deal with real instead binary input vectors, ART-3 which incorporates medium-term memory (MTM) and fuzzy-ART which includes fuzziness concepts (see Sect. 3.3) (Carpenter and Goldberg 2009).

Other interesting paradigm which uses unsupervised learning are the self-organizing maps (SOM), also known as Kohonen networks, after Teuvo Kohonen who proposed them in the eighties. It is especially convenient for visualizing high-dimensional data. SOM's are frequently used for solving tasks involving clustering, classification and monitoring.

The architecture of SOM's (Fig. 3.15) is composed by two layers: an m-neuron input layer, where m is the dimensionality of the input data vectors, and an n^2-neuron output layer, arranged in and $n \times n$ matrix. Neurons in the input layer are fully connected with neurons in the output layer by the matrix **W**, where $w_{i,j}$ is the weight connecting the ith input neuron with the jth output neuron.

Two important concepts play an important role in the training process of a SOM. The first one is the weight adaptation process and the second one is the idea of a topological neighborhood of neurons.

The weights are updated by following the expression (Marini et al. 2005):

$$w_{i,j}^{(t+1)} = w_{i,j}^{(t)} + \eta \left(1 - \frac{d_j}{d_{\max} + 1} \right) \left(X_i - w_{i,j}^{(t)} \right) \tag{3.20}$$

where $w_{i,j}^{(t+1)}$ and $w_{i,j}^{(t)}$ are the weight connecting the ith input with the jth neuron, at iterations t and $t + 1$, respectively; η, is the learning rate; d_j, is the topological distance from the jth neuron to the winning neuron (i.e., the neuron having the

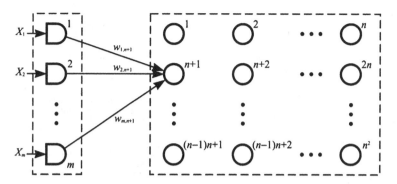

Fig. 3.15 Architecture of a SOM

shortest distance to the current input); $d_{\max}$, is the size of neighborhood; and X_i is the value of the ith component of the current input vector.

For a two dimensional arrangement of output neurons, the topological distance between two neurons, located at positions (r_1, c_1) and (r_2, c_2), respectively (being r the row and c the column), is defined by:

$$d = \left\lfloor \sqrt{(r_1 + r_2)^2 + (c_1 - c_2)^2} \right\rfloor; \tag{3.21}$$

where $\lfloor \bullet \rfloor$ denotes the floor function, i.e., the greater integer number less than or equal to the argument.

The size of neighborhood is initially defined covering the entire network, decreases through the learning process, usually by the exponential function. Thence, for iteration t (Ghaseminezhad and Karami 2011):

$$d_{\max}^{(t)} = d_{\max}^{(0)} \exp\left(-\frac{t}{NT}\right); \tag{3.22}$$

where $d_{\max}^{(0)}$ is the initial neighborhood size; T is the total iterations; and N is a value (usually $N \approx 3$).

The value of learning rate also decreases through the learning process, by following the expression (Marini et al. 2005):

$$\eta^{(t)} = \left(\eta^{(0)} - \eta^{(\mathrm{end})}\right)\left(1 - \frac{t}{T}\right) + \eta^{(\mathrm{end})}; \tag{3.23}$$

where $\eta^{(0)}$ are $\eta^{(\mathrm{end})}$ the initial and the ending learning rates, respectively.

Sometimes, other expressions are used for updating the neighborhood size and the learning rate, but the central idea remains constant. In Fig. 3.16, it is shown the block diagram of the learning algorithm for a SOM.

Fig. 3.16 Training algorithm
of the SOM

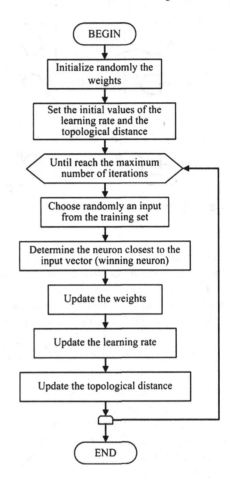

3.2.7 Warnings and Shortcomings in the Use of Neural Networks

Although being a powerful and useful modeling tool, neural networks have been repeatedly misused. Perhaps the main fault on using neural networks is, precisely, to use them without having a solid theoretical background on their principles and functioning. Often, neural networks are called "black boxes". This term is correct in the sense of they are empirical model which relate input and output data without considering any causal or phenomenological relationship between them. So, they are no more "black box" than any other empirical model, such as a linear regression. "Black box" does not mean than the internal functioning of the neural network is not known.

The most frequent mistake, as reported in the literature, is the use of too few data for training complex networks (Sha and Edwards 2007). In spite of any other biological or structural consideration, neural networks can be seen as a special case of non-linear regression expression, where weights and biases are the free

Fig. 3.17 Mathematically indetermination

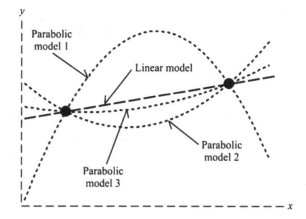

parameters that should be adjusted during the training process. Therefore, it is not mathematically possible to determine the value of these parameters if there are, at least, as many training samples as the total number of parameters to be computed.

For example, in a fully connected MLP, with a 3-6-1 structure (i.e., 3 inputs, 6 nodes in the hidden layer and 1 output), there are $3 \times 6 + 6 \times 1 = 24$ weights and $3 + 6 = 9$ biases, making a total of 33 parameters to be computed. Consequently, if less than 33 training samples are used with this purpose, the training process will be mathematically undetermined.

To illustrate this problem, see Fig. 3.17. If a linear model, with two free parameters, is fitted from two points (black circles) a unique model will be obtained (dashed line) and the process is perfectly defined. On the contrary, is a parabolic model, having three free parameters is fitted from these two points, infinite different parabolas (dotted lines) can be obtained passing through them; so, the process is mathematically undetermined.

Furthermore, if the amount of training data is exactly equal to the number of free parameters, they can be computed, however, the statistical significance of the computed values has no sense. It is necessary to use much more training samples to increase the statistical significance of the training process results.

Other common mistake arises when complex neural networks are used to fit data with simple trends in their relationships. In this situation, the model fits perfectly the training data. Apparently this gives us with a very good model, but actually it is not the case. Experimental data always has some error components (usually known as noise). If the proper model is selected (see dashed line, Fig. 3.18), it will describe the behavior of the data in spite of the presence of noise. On the contrary, an over-fitted model (see dotted line, Fig. 3.18) will fit even the noise, so the generalization capabilities of this model are very low.

In order to avoid the over-fitting problem in neural network training, the predictions of the fitted network must be compared not only with the observed values of the training set, but also with the observed values of some validation set which had not been used in the training processes. If there are not statistically

Fig. 3.18 Over-fitting

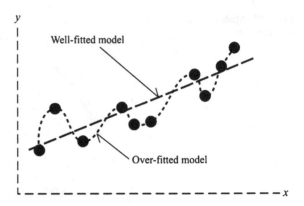

significant differences between the residuals of both sets, then there is no indication of over-fitting.

Another important good practice in dealing with neural networks, often ignored, is to compute multiples runs in order to avoid random influences, for example, from the weights initialization, or the sequence of the training data (Flexer 1996). Moreover, when some parameters of the network, such the number of neurons in the hidden layers or the learning rate, are tuned for obtaining the best network performance, a third independent data set must be used to test the optimized network after the training process.

Finally, it must be pointed the need of perform statistic test, such as the t-test, for comparing the performance of different networks (Flexer 1996).

3.3 Fuzzy Logic

Fuzzy logic is a branch of the soft computing aimed to deal with the uncertainty of the real world. It was introduced by L. A. Zadeh of University of California, Berkeley, U.S.A., in the 1960 years, as a means for modeling the uncertainty of natural language. (Ramík 2001).

Although probability theory also deals with uncertainty, both proceed in basically different ways: while probability considers that parameters have exact crisp values, but these values are unknown, fuzzy logic considers that parameters have intrinsically fuzzy values. Therefore, fuzzy logic deals with the vagueness of the real world while probability deals with our ignorance on this world.

In classic Boolean logic, proposition can only have two possible values: true or false. On the contrary, in fuzzy logic, propositions can have different values that are known as degree-of-true, ranging between 0 and 1. A fuzzy set (or a fuzzy subset) can be defined as a set of ordered pairs, each with the first element from certain set S, and the second element from the interval [0, 1], and having exactly one ordered pair for each element of S (Ramík 2001).

Fig. 3.19 Membership functions

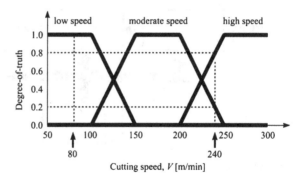

The truth of any proposition if given by fuzzy subsets called membership functions, which establish the levels of degree-of-truth corresponding to the values of the parameter. For example (see Fig. 3.19), three different membership functions can be defined for the cutting speed in certain machining process, defining the categories "low", "moderate" and "high", as the graph shows.

Then, for a cutting speed value of 80 m/min, the proposition "cutting speed is low" has a degree-of-truth equal to one (i.e., cutting speed is completely low), while the propositions "cutting speed is moderate" and "cutting speed is high" have degree-of-truth of zero (i.e., cutting speed is not moderate or low at all). In other case, for a cutting speed of 240 m/min, the proposition "cutting speed is low" has a degree-of-truth of zero (i.e., cutting speed is not low at all), "cutting speed is moderate" has a degree-of-truth of 0.2 (i.e., cutting speed is slightly moderate) and "cutting speed is high" has a degree-of-true of 0.8 (i.e., cutting speed is fairly high).

For two fuzzy propositions A y B, the following operations are defined (Ramík 2001):

$$
\left.
\begin{array}{lll}
\text{NOT operation:} & \text{truth}(\neg A) = 1 - \text{truth}(A) \\
\text{AND operation:} & \text{truth}(A \wedge B) = \min\{\text{truth}(A), \text{truth}(B)\} \\
\text{OR operation:} & \text{truth}(A \vee B) = \max\{\text{truth}(A), \text{truth}(B)\}
\end{array}
\right\}.
\qquad (3.24)
$$

By considering the membership functions defined in Fig. 3.19, for a cutting speed of 240 m/min, the following statements will result:

$$\text{truth}\{\neg(V \text{ is } "high")\} = 1.0 - 0.8 = 0.2$$

$$\text{truth}\{(V \text{ is } "high") \wedge (V \text{ is } "moderate")\} = \min(0.8, 0.2) = 0.2$$

$$\text{truth}\{(V \text{ is } "high") \vee (V \text{ is } "moderate")\} = \max(0.8, 0.2) = 0.8$$

Now, combining the before defined operations, fuzzy inference rules can be formed, in an IF–THEN form (Ross 2004):

$$\text{IF } premise \text{ THEN } conclusion \qquad (3.25)$$

where the rule's premise (between IF and THEN) describes to what degree the rule applies, while the rule's conclusion (following THEN) assigns a membership function to each of one or more output variables.

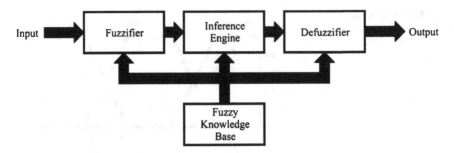

Fig. 3.20 Fuzzy inference system

For example, by considering three fuzzy variables: cutting speed, V, feed, f, and cutting force, F_C, for a machining processes, with their respective membership functions "low", "moderate" and "high", these variables can be related by the following inference rule:

$$\text{IF } (V \text{ is } "high") \wedge (f \text{ is } "low") \text{ THEN } (F_C \text{ is } "low").$$

A fuzzy inference system (FIS), also known as fuzzy rule-based system, fuzzy associative memory or fuzzy logic controller, is a set of fuzzy inference rules, arranged in certain form and working together to produce a mapping from a given input to an output. A FIS is composed by three main parts (see Fig. 3.20): a fuzzifier that converts the crisp input into a fuzzy variable by using a set of membership functions; an inference engine that transform the fuzzy input into a fuzzy output through a set of inference rules; and a defuzzifier that convert back the fuzzy output to a crisp value, by using membership functions. All the membership functions and fuzzy inference rule conform the fuzzy knowledge base (Siler and Buckley 2005).

There are two common FIS's: Mamdani's and Sugeno's inference systems. Mamdani FIS is, probably the most used in implementing fuzzy systems. It uses a set of and-related fuzzy inference rules and, combine the result of the different rules by using the maximum of the obtained fuzzy sets (see Fig. 3.21).

The combined fuzzy set is finally defuzzified, usually, by computing the centroid of the area, although sometimes other parameters are used for this function, such as the bisector of the area or the smaller, mean or larger values of the maximum (see Fig. 3.22).

The Sugeno FIS, also called TSK after Takagi, Sugeno and Kang who propose it in the middle eighties, uses or-based inference rules, instead of and-based ones Fig. 3.23. Each rule gives a crisp outcome, which can be seen as the firing weight of the rule. Additionally, it gives an output, computed from the crisp values of the inputs, usually by using a polynomial function.

Finally the overall output is computed by performing a weighted sum of the individual outputs:

$$z = \frac{\sum_i w_i x_i}{\sum_i w_i}. \tag{3.26}$$

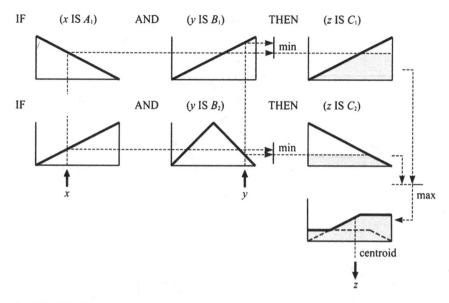

IF   (x IS A_1)   AND   (y IS B_1)   THEN   (z IS C_1)

IF   AND   (y IS B_2)   THEN   (z IS C_2)

Fig. 3.21 Mamdani fuzzy inference system

Fig. 3.22 Defuzzification methods

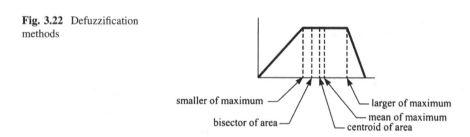

smaller of maximum

bisector of area

larger of maximum

mean of maximum

centroid of area

3.4 Neuro-Fuzzy Systems

By combining the advantages of artificial neural networks and fuzzy logic it is possible to build more powerful and versatile systems. Neuro-fuzzy systems can combine both tools in different ways (Fullér 2000). The two basic approaches are shown in Fig. 3.24.

In the first approach (Fig. 3.24a), several neural networks take the input and process it. Then, the outcomes of the networks are combined by a fuzzy inference system, which gives the final output. The second approach (Fig. 3.24b) is very similar: several fuzzy inference systems take the input and return intermediate outcomes. After that, an artificial neural network combines theses outcomes and gives the output.

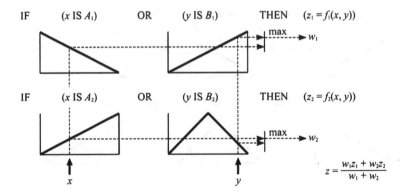

Fig. 3.23 Sugeno fuzzy inference system

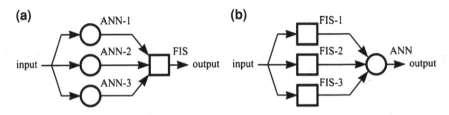

Fig. 3.24 Basic neuro-fuzzy systems

Fig. 3.25 Neuro-fuzzy system

Other more complex approaches are also possible. For example, a neural network can be used for tuning up the parameters of a fuzzy inference system (Fig. 3.25). The output of the inference system together with the input is used as the training set of the neural network.

One of the Most Recently Used Neuro-Fuzzy Approach is the Adaptive Neuro-Fuzzy Inference System (ANFIS), which is a nultilayer feed-forward network using learning algorithms and fuzzy reasoning to map the relationship between the input and output variables.

An ANFIS is composed by five layers (Fig. 3.26) (Jang 1993). The first one is the fuzzification layer. It takes the N inputs, $x_1, \ldots, x_N$, and applies the membership

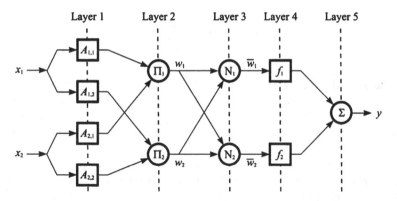

Fig. 3.26 ANFIS architecture

functions, $A_{i,j}$, where $j = 1, \ldots, M$ (M is the number of used fuzzy sets). Typical membership functions used in this layer have the form:

$$A_{i,j}(x_i) = \frac{1}{1 + \left|\frac{x_i - c}{a}\right|^{2b}};$$

(3.27)

where a, b and c are referred to as premise parameters and are adaptively tuned up during the training process.

The nodes of the second layer represent the firing strength of the rule, which is obtained by computing the product of the incoming signals:

$$O_{2,j} = w_j = \prod_{i=1}^{N} A_{i,j}(x_i).$$

(3.28)

The third layer normalizes the output of the second one. These values, which are also known as normalized firing strengths, can be computed trough the expression:

$$O_{3,j} = \bar{w}_j = \frac{w_j}{\sum_{i=1}^{M} w_i}.$$

(3.29)

The nodes of the fourth layer have an output that can be computed as:

$$O_{4,j} = \bar{w}_j f_j = \bar{w}_j \left(q_j + \sum_{i=1}^{N} p_{i,j} x_i \right);$$

(3.30)

where q_j and $p_{i,j}$ are the adaptive consequent parameters.

Finally, the fifth layer computes its output, which is also the output of the whole network, by the sum of all the outputs of the fourth layer nodes:

$$O_5 = \sum_{j=1}^{M} O_{4,j} = \sum_{j=1}^{M} \bar{w}_j f_j. \tag{3.31}$$

The training process of an ANFIS is carried out in two steps. In the first forward step, the algorithm uses least-squares method to identify the consequent parameters on the layer 4. Then, in the second backward step, the errors are propagated backward and the premise parameters are updated by gradient descent. The process is iteratively repeated with a set of training samples until reach some prescribed error value (Po 2002).

When the number of rules is not restricted, an ANFIS has unlimited approximation power for matching any nonlinear function arbitrarily well on a compact set (Zhang et al. 2011).

3.5 Metaheuristic Optimization

3.5.1 Optimization Basis

Optimization plays a key role in science and technology braches because it is closely related with technical efficacy and economical efficiency. Especially in the study of manufacturing processes, optimization has been an essential topic through the years and, even now, it remains being a very important research field (Chandrasekaran et al. 2010).

Optimization can be classified in single-objective, when only one target must be optimized, and multi-objective, when several objectives must be considered simultaneously.

A single-objective optimization problem can be defined as the task of find the value of the decision variable $\mathbf{x} = \{x_1, x_2, \ldots, x_M\}^{\mathrm{T}}$, which makes minimum some target function,

$$y = y(\mathbf{x}); \tag{3.32}$$

while satisfies the set of inequality constrains,

$$g_i = g_i(\mathbf{x}) \leq 0, \quad i = 1, \ldots, P; \tag{3.33a}$$

and the set of equality constrains,

$$h_i = h_i(\mathbf{x}) = 0, \quad i = 1, \ldots, Q. \tag{3.33b}$$

It must be noted than the fact that the target must be minimized is purely conventional. Any minimization target function may be transformed to a maximization one just by multiplying it by minus one.

Definition of a multi-objective optimization problem can be made in a similar way, only considering a vector target function,

Fig. 3.27 Paretian
dominance

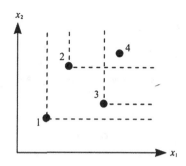

$$\mathbf{y} = \mathbf{y}(\mathbf{x}); \tag{3.34}$$

instead a scalar one. However, there is a problem because the comparison between
two vectors is not defined.

There are three main approaches for dealing with this situation (Van Veldhuizen
and Lamont 2000):

- *A priori* approach: The different targets are combined into a single one, turning
  the multi-objective problem into a single-objective one before the actual optimization process takes place. This approach can be carried out by using different
  techniques such as linear or nonlinear combination, aggregation by ordering,
  etc.
- Progressive approach: Decision making and optimization are intertwined. Partial preference information is provided upon which optimization occurs, providing an updated set of solutions for the decision-maker to consider.
- *A posteriori* approach: A set of optimal candidate solutions is obtained though
  the optimization process. After that, the decision-maker choose the most convenient solution for the specific conditions.

Currently, a *posteriori* approach is considered to be the most versatile technique
and has received much attention from the researchers. For using this approach, the
Paretian dominance criterion must be used. A vector, $\mathbf{x} = \{x_1, x_2, \ldots, x_k\}$ is said
to dominate other vector, $\mathbf{y} = \{y_1, y_2, \ldots, y_k\}$ if and only if $x_i \leq y_i$, for all
$i \in \{1, \ldots, k\}$, and there is at least one $i \in \{1, \ldots, k\}$ such that $x_i < y_i$. In Fig. 3.27
it is graphically represented the concept of Paretian dominance for two-dimen-
sional vector. Here, vector 1 dominates all the other vectors, while vectors 2 and 3
dominate vector 4. Neither vector 2 dominates vector 3 nor vector 3 dominate
vector 2. Vector 4 does not dominate other vector.

Consequently, a non-dominated vector is that which does not have other vector
within certain set which dominates. In Fig. 3.28, black circles represent non-
dominated vectors. Conversely, white circles represent dominated vectors.

In a multi-objective optimization problem, the set of non-dominated solutions,
which simultaneously belong to the feasible region (i.e., that fulfill all the constraints) is called Pareto optimal and its corresponding target vectors, Pareto front

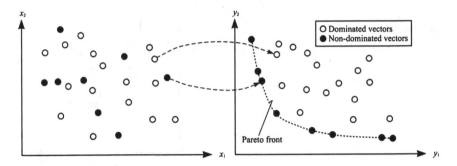

Fig. 3.28 Non-dominated and dominated vectors

(see Fig. 3.28). These solutions are optimal in the wide sense that there is no other solution in the feasible region that improves one optimization target without worsens some other.

Therefore, the solution of this optimization problem by using the a posteriori approach consist in obtaining the Pareto's front and, after that, select the most proper solution, form this front, considering the conditions.

Although there are numerical methods for solving multi-objective optimization problems, they do not allow obtaining the Pareto's front in a single run. Also they also cannot effectively deal with problems involving discontinuity, non-differentiability or multimodality in the target functions.

On the contrary, the so-called heuristic or stochastic optimization methods, which are optimizations techniques inspired in some human or natural processes, although without a solid mathematical foundation, have proved their ability for finding optimal or near-optimal solution sets in many complex optimization problems (Schneider and Kirkpatrick 2006). These methods have been widely applied for solving manufacturing optimization problems through the last years (Chandrasekaran et al. 2010).

3.5.2 Evolutionary Computation

Evolutionary computation is the field of soft-computing that studies computational system by using ideas inspired from natural evolution (Sarker et al. 2003). They include several branches, which differ mainly if their representation of potential solutions and, consequently, in the operators and algorithms they use for solving the optimization problems.

The two main approaches in evolutionary optimization are the evolution strategies (ES), also known as the German school, proposed by Rechenberg and Schwefel in 1965 and the genetic algorithms (GA), developed by Holland in the seventies, and commonly known as the American school.

Fig. 3.29 Basic algorithm
for ES's and GA's

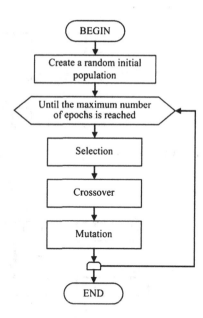

ES's uses real numbers for coding the values of the decision variable of the potential solutions. This approach requires a simpler implementation and works better with simple problem involving continuous numeric variables. On the contrary, GA's use a string (often, a bitstring) for this purpose. It requires most elaborated operators but it is more robust (i.e., it can be indistinctly applied to problems involving not only continuous but also discrete and even qualitative variables). Genetic algorithms are suitable for solving combinatorial optimization problems (Schneider and Kirkpatrick 2006).

In spite of their differences, both approach share several features. First at all, they use a set of candidate solutions (called solution population or, simply, population) for carrying out a parallel search. Moreover, they imitate the natural evolution by implementing three basic operators: selection, crossover and mutation. The basic algorithm for these approaches (see Fig. 3.29) consists in the generation of a random initial population followed by a loop where selection, crossover and mutation are successively applied to the current population for generating a new one which hopefully will be closer to the optimum.

Selection of the individual that will take part in the generation of the new population is one of the most important operators in any ES or GA. It can be faced in different ways, including different kinds of ranking, were individual are sorted considering some fitness criterion (which match with the optimization targets) and the probably of been selected is determined by the position in the ranking and tournament, where a couple of individuals are randomly chosen and compared for including the winner into the parents of the new population. Of course, a lot of more complex approaches are possible and have been already proposed.

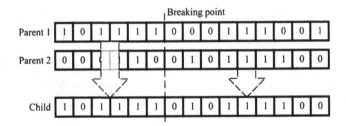

Fig. 3.30 Single breaking point GA crossover

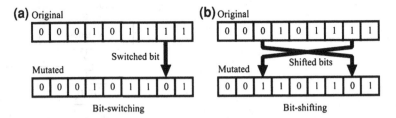

Fig. 3.31 Mutation in GA's

Through the crossover operator new solutions (children individuals) are obtained from current solutions (parent individuals), so that the characteristics of the new solutions are a combination of those of their parents. In ES's the children get some intermediate value of the real number representing the characteristics of their parents. In GA's the crossover is inspired in the breaking and recombination of chromatides in natural organisms (see Fig. 3.30).

Mutation allows random small changes is the individuals through the optimization process. In ES, mutation takes place simply by adding some random number to the current value of the decision variable. In GA, mutation can be carried out by changing the value of some bits (bit-switching, see Fig. 3.31a) or by moving some bits to other positions into the bitstring (bit-shifting, see Fig. 3.31b). The value of the mutation likelihood must be kept low to avoid the method degeneration in a simple random search.

3.5.3 Evolutionary Multi-Objective Optimization

Evolutionary computation has been widely applied for solving multi-objective optimization problems. However, the application of this technique to such a field is not straight-forward as it involves several important challengers.

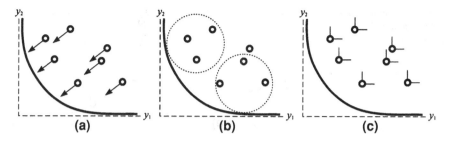

Fig. 3.32 Schematic representation of fitness assignment approaches

The first problem, in dealing with multi-objective optimization is the inexistence of a single target. As there is not a comparison criterion for vector, the optimization targets and the fitness function are not identical.

There are three main approaches for dealing with fitness assignment in evolutionary multi-objective optimization (Zitzler et al. 2004):

- Aggregation-based approach (see Fig. 3.32a) combines the different target into a single parameterized fitness function. The parameters are changed through the optimization run in order to obtain the Pareto front.
- Criterion-based approach (see Fig. 3.32b) switches between the objectives during the selection phase. It can be done by using different criteria when selecting the parents for each generation, or by changing the criterion from generation to generation.
- Pareto dominance-based approach (see Fig. 3.32c) uses the concept of non-dominated vector. Individuals can be sorted by considering the number of individual by which an individual is dominated (dominance rank). Other approach is to divide the population in several successive fronts and the fitness is established according the front containing the individual (dominance depth). A third way if to consider the number of individual which are dominated by the individual (dominance count).

A second issue than must be always kept in mind in evolutionary multi-objective optimization is the preservation of the diversity. If some techniques are not incorporated to the algorithm for dealing with this problem, the obtained solution can be concentrated only in a small sector (see Fig. 3.33a) instead of being regularly distribute through all the Pareto front (Fig. 3.33b).

The preservation of the diversity can be guaranteed with the incorporation of individual density information in the selection process. This can be implemented in three approaches (Zitzler et al. 2004):

- Kernel methods (see Fig. 3.34a) estimate the population density by using a so-called Kernel function, which takes the distances from a given point to the others as argument. The most commonly used Kernel method in evolutionary optimization is the fitness shearing technique.

Fig. 3.33 Pareto fronts without and with diversity preservation

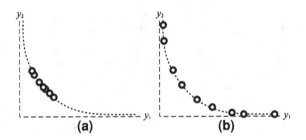

- Nearest neighbor techniques (see Fig. 3.34b) takes the distance from a given point to its kth nearest neighbor. Usually, the density estimator is the inverse of this distance.
- Histograms (see Fig. 3.34c) use a hyper-grid to define the neighborhood within the space, estimating the population density around a given individual as the number of individual in the same box of the hyper-grid.

Diversity preservation techniques can be implemented by considering the density distribution in the decision variable space or in the target space, although the last variant if the often used in evolutionary multi-objective optimization.

Elitism avoid the losing the best individual during the optimization run. Elitism can be carried out in two ways:

- Without using an external memory archive (see Fig. 3.35a), a offspring is created from the current population. After that, the new population is created with the best individual from both the current population and the offspring.
- With the use of an external memory archive (see Fig. 3.35b), the new population is created directly from the current one. Then, the external archive is updated by adding the non-dominated solutions from the new population. Usually, after each update, the external archive is filtering for removing the non-dominated solutions and for avoiding the excessive increment of the used memory.

Several evolutionary multi-objective algorithms have been proposed. The pioneer in this field was the vector evaluated genetic algorithm (VEGA) proposed by Schaffer in the middle eighties. VEGA consists in a simple GA with a aggregation-based selection mechanism, which switches between the objective through the optimization process.

A so-called first generation, developed in the nineties, typically adopted niching or fitness shearing to overcome the shortcoming of VEGA (Coello 2003). Some representative approaches of this group were the non-dominated sorting genetic algorithm (NSGA), niched-Pareto genetic algorithm (NPGA) and multi-objective genetic algorithm (MOGA).

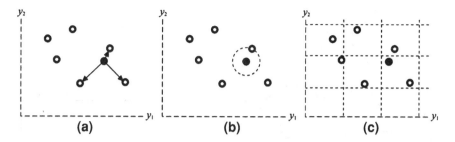

Fig. 3.34 Diversity preservation approaches

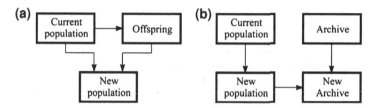

Fig. 3.35 Elitism techniques

The second generation raised in the first decade of the XXI century, by the inclusion of the concept of elitism. It included the NPGA-II, NSGA-II and the micro genetic algorithm (μGA) (López and Coello 2009).

3.5.4 Swarm Intelligence

The term swarm intelligence encloses a group of techniques, mainly used for optimization purposes, which are based in the behavior of groups of individuals. This paradigm is supported by the idea that the ability of the whole community for solving some problems if greater than the ability of a separated member. The way ants find the most convenient path between two points is an example of this capability.

Several swarm intelligence based techniques have been proposed in the last years. Nevertheless, the most used approaches in manufacturing processes optimization are the ant-colony optimization (ACO), the particle swarm optimization (PSO) and the simulated annealing (SA) which will be briefly described below.

Ant-colony optimization (ACO) is a population-based optimization algorithm, proposed by Dorigo and coworkers in the beginning nineties of the last century, which is inspired in the foraging behavior of real ants (Bonabeau et al. 1999). When ants find a food source, they travel between this source to the nest initially following random paths. While traveling, they deposit a chemical substance called pheromone, which attracts the other ants. As the shortest path is travelled more

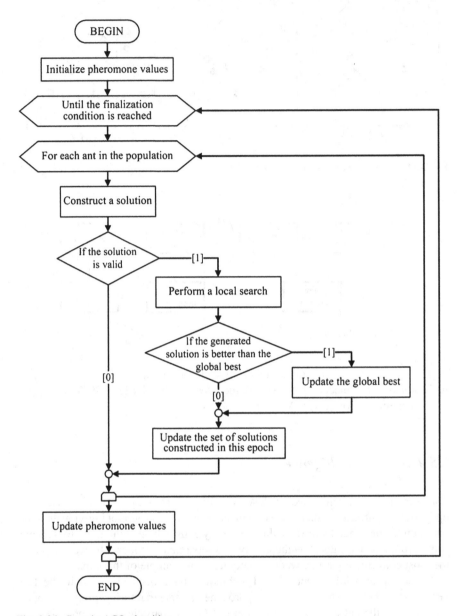

Fig. 3.36 Generic ACO algorithm

times it receives a greatest amount of pheromone and, therefore, attracts more ants. On the contrary, as pheromone evaporates, longer paths lose the attractiveness through the time. Finally, only the shorter path is travelled by all the ants (Ghaiebi and Solimanpur 2007).

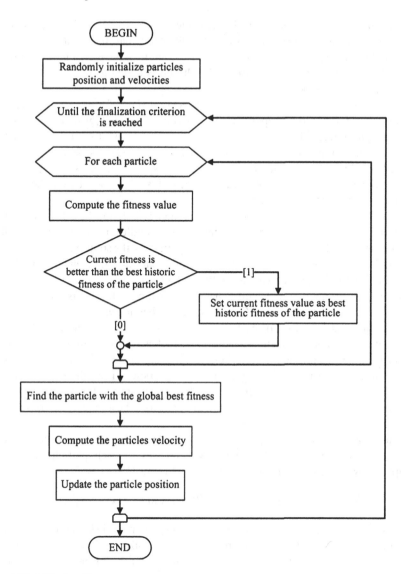

Fig. 3.37 PSO algorithm

The algorithm for a generic ACO (see Fig. 3.36) (Dorigo and Blum 2005) begins with the initialization of all the pheromone values at some value greater than zero. Then, a loop is performed until some finalization condition is reached. Inside this loop, a second iteration takes places where, for each ant in the population, a solution is created probabilistically by adding elements from the finite set of solution components. The probability for the choice of one solution component depends on the corresponding pheromone values.

Then, if the created solution is valid (feasible) a local search, which improves the algorithm's overall performance, is carried out. After that, the created solution is compared with the global best solution. If it is better, then the global best solution is updated to the new value. Also, the new solution is added to the set of solutions constructed in this epoch.

After the inner loop, the pheromones values are updated for increasing the pheromone values on solution components that have been found in high quality solutions.

Although conceived for solving combinatory problems, a modification of the ACO, called continuous ant colony optimization (CACO) has been proposed. It divided the domain of the target function into a specific number of regions and then it uses a combination of local and global search for finding the global optimum (Saravanan et al. 2005).

Particle swarm optimization (PSO) is another population based optimization technique presented by Eberhart and Kennedy in the middle nineties. It is based on the social behavior of bird flocking or fish schooling. In PSO, the solutions move through the problem space following the current optimum particle (Saravanan et al. 2005). Each solution (called particle) has a position, which is actually the vector of decision variables, and a velocity representing the rate of change of this decision variables.

The basic algorithm of the PSO (see Fig. 3.37) begins with the initialization of the positions and velocities of each particle at some random values. Then a loop take place until some finalization condition is reaches (usually a maximum number of epochs or some minimum error criterion). In every epoch of the loop, the fitness value (target function) is computed for all the particles. Then the current fitness value for each particle, $y_i^{(t)}$, is compared with the historic best fitness value for this particle, y_i^*. If the current fitness is best than the historic best fitness, this is updated with the current value.

After evaluating all the population, the particle having the global best fitness value, y^{**}, is determined. Then the new velocities of each particle, $v_i^{(t+1)}$ are computed by using the following expression (Noorul Haq et al. 2006):

$$\mathbf{v}_i^{(t+1)} = w\mathbf{v}_i^{(t)} + c_1 \text{rand}_{(0,1)}(\mathbf{x}_i^* - \mathbf{x}_i^{(t)}) + c_2 \text{rand}_{(0,1)}(\mathbf{x}^{**} - \mathbf{x}_i^{(t)}). \qquad (3.35)$$

where $\mathbf{x}_i^{(t)}$ and $\mathbf{v}_i^{(t)}$ are the position and velocity for the ith particle in the current epoch; w is the so-called inertia term, representing the relative weight of the previous velocities; $\text{rand}_{(0,1)}$ is a uniformly distributed random number in the interval $(0, 1)$; $\mathbf{x}_i^*$ is the historic best position for the ith particle; $\mathbf{x}^{**}$ is the global best position; and c_1 and c_2 are learning parameters, usually selected in the interval $[0, 4]$. Then, the position of every particle is updated by doing (Noorul Haq et al. 2006):

$$\mathbf{x}_i^{(t+1)} = \mathbf{x}_i^{(t)} + \mathbf{v}_i^{(t)}. \qquad (3.36)$$

Several variations of the PSO have been proposed, differing in the initialization techniques or in the selection of the values for the learning parameters. Moreover, more complex algorithms, even dealing with multi-objective optimization, can be found in the literature (Attea 2010).

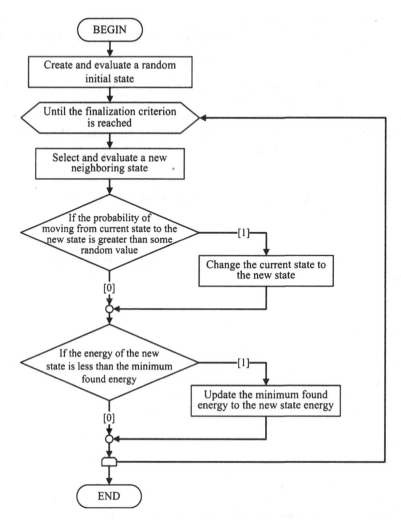

Fig. 3.38 SA algorithm

Simulated annealing (SA) is another frequently used metaheuristic optimization technique. It resembles the cooling process of a molten metal through annealing (Saravanan et al. 2005). The method was independently introduced by Kirkpatrick and co-workers and by Černý in the middle eighties (Bertsimas and Tsitsiklis 1993). In SA each point in the decision variable space can be seen as the state of some physical system, and the optimization target as the corresponding energy; therefore, the optimization process is similar to process of achieving a minimum energy state (equilibrium).

The basic idea in the SA is to iteratively consider neighboring states of the current state and probabilistically decide either moving the system to the new state or staying in the current state. The process is repeated until some equilibrium state is reached (see Fig. 3.38).

The neighbor selection depends on the specific problem. For continuous decision variables it can be done by using the Gaussian distribution.

Several expressions have been proposed for computing the probability of changing from the current state, $\mathbf{x}$, to a new state, $\mathbf{x}'$, but perhaps the most commonly used is the Metropolis criterion (Schneider and Kirkpatrick 2006):

$$p(\mathbf{x} \rightarrow \mathbf{x}') = \begin{cases} \exp\left(-\frac{y(\mathbf{x}')-y(\mathbf{x})}{T}\right) & \text{if } y(\mathbf{x}') - y(\mathbf{x}) > 0 \\ 1 & \text{otherwise} \end{cases};\qquad(3.37)$$

where $y(\mathbf{x})$ is the target function evaluated at state $\mathbf{x}$, and T is a parameter called temperature (by an analogy with the physical annealing) which decreases with the time.

References

B.A. Attea, A fuzzy multi-objective particle swarm optimization for effective data clustering. Memetic Comp. **2**, 305–312 (2010). doi:10.1007/s12293-010-0047-2

L. Behera, S. Kumar, A. Patnaik, On adaptive learning rate that guarantees convergence in feedforward networks. IEEE T Neural Netw. **17**, 1116–1125 (2006). doi:10.1109/TNN.2006.878121

T. Bertsimas, D. Tsitsiklis, Simulated annealing. Stat. Sci. **8**, 10–15 (1993). doi:10.1214/ss/1177011077

E. Bonabeau, M. Dorigo, G. Theraulaz, *Swarm Intelligence: From Natural to Artificial Systems* (Oxford University Press, New York, 1999)

M.D. Buhmann, *Radial Basis Functions* (Cambridge University Press, Cambridge, 2004)

G.A. Carpenter, S. Goldberg, Adaptive Resonance Theory, in *Encyclopedia of Machine Learning*, ed. by C. Sammut, G.I. Webb (Springer, New York, 2009)

M. Chandrasekaran, M. Muralidhar, C. Murali Krishna, U.S. Dixit, Application of soft computing techniques in machining performance prediction and optimization: a literature review. Int. J. Adv. Manuf. Tech. **46**, 445–464 (2010). doi:10.1007/s00170-009-2104-x

C.A. Coello, Multiobjective Optimization: Current and Future Challenges, in *Advances in Soft Computing: Engineering, Design and Manufacturing*, ed. by J. Benitez, O. Cordon, F. Hoffmann, R. Roy (Springer, London, 2003)

M. Dorigo, C. Blum, Ant colony optimization theory: A survey. Lect. Notes Comput. Sci. **344**, 243–278 (2005). doi:10.1016/j.tcs.2005.05.020

Flexer A Statistical evaluation of neural network experiments: minimum requeriments and current practices. In: Trappl R (ed.) 13th European Meeting on Cybernetics and Systems Research, Vienna (1996)

R. Fullér, *Introduction to neuro-fuzzy systems* (Physica-Verlag, Heidelberg, 2000)

H. Ghaiebi, M. Solimanpur, An ant algorithm for optimization of hole-making operations. Comput. Ind. Eng. **52**, 308–319 (2007). doi:10.1016/j.cie.2007.01.001

M.H. Ghaseminezhad, A. Karami, A novel self-organizing map (SOM) neural network for discrete groups of data clustering. Appl. Soft. Comput. **11**, 3771–3778 (2011). doi:10.1016/j.asoc.2011.02.009

M.T. Hagan, H.B. Demuth, M. Beale, *Neural Network Design* (China Machine Press, Beijing, 2002)

R.L. Harvey, *Neural Network Principles* (Prentice Hall, Englewood Cliffs, 1994)

L.G. Heins, D.R. Tauirtz, *Adaptive Resonance Theory (ART): An Introduction Internal Report 95–35* (Department of Computer Science, Leiden University, Leiden, 1995)

Y.H. Hu, J.-N. Hwang, *Handbook of Neural Network Signal Processing* (CRC Press, Boca Raton, 2002)

J.S.R. Jang, ANFIS: Adaptive-network-based fuzzy inference system. IEEE T Syst. Man Cyb. **23**, 665–685 (1993). doi:10.1109/21.256541

N.K. Kasabov, *Foundations of Neural Networks, Fuzzy Systems and Knowledge Engineering* (MIT Press, Cambridge, 1998)

A. López, C.A. Coello, Multi-objective evolutionary algorithms: a review of the state-of-the-art and some of their applications in chemical engineering, in *Multi-Objective Optimization Techniques and Applications in Chemical Engineering*, ed. by R.G. Pandu (World Scientific, Singapore, 2009)

F. Marini, J. Zupan, A.L. Magri, Class-modeling using Kohonen artificial neural networks. Anal. Chim. Acta **544**, 306–314 (2005). doi:10.1016/j.aca.2004.12.026

A. Noorul Haq, K. Sivakumar, R. Saravanan, K. Karthikeyan, Particle swarm optimization (PSO) algorithm for optimal machining allocation of clutch assembly. Int. J. Adv. Manuf. Tech. **27**, 865–869 (2006). doi:10.1007/s00170-004-2274-5

S.P. Po, The application of an ANFIS and grey system method in turning tool-failure detection. Int. J. Adv. Manuf. Tech. **19**, 564–572 (2002). doi:10.1007/s001700200061

S. Rajasekaran, G.A. Vijayalakshmi Pal, *Neural Networks, Fuzzy Logic and Genetic Algorithms: Synthesis and Applications* (Prentice-Hall of India, New Delhi, 2003)

J. Ramík, Soft computing: Overview and Recent Developments in Fuzzy Optimization. University of Ostrava (Czech Republic) (2001), http://ac030.osu.cz/irafm/ps/softco01.pdf

T.J. Ross, *Fuzzy Logic With Engineering Applications* (Wiley, West Sussex, 2004)

R. Rojas, *Neural Networks: A Systematic Introduction* (Springer, Berlin, 1996)

G. Roussos, B.J.C. Baxter, Rapid evaluation of radia basis functions. J. Comput. Appl. Math. **180**, 51–70 (2005). doi:10.1016/j.cam.2004.10.002

R. Saravanan, R.S. Sankar, P. Asokan, K. Vijayakumar, G. Prabhaharan, Optimization of cutting conditions during continuous finished profile machining using non-traditional techniques. Int. J. Adv. Manuf. Tech. **26**, 30–40 (2005). doi:10.1007/s00170-003-1938-x

R. Sarker, M. Mohammadian, X. Yao, *Evolutionary Optimization* (Kluwer Academic Publishers, New York, 2003)

W. Sha, K.L. Edwards, The use of artificial neural networks in materials science based research. Mater Design **28**, 1747–1752 (2007). doi:10.1016/j.matdes.2007.02.009

W. Siler, J.J. Buckley, *Fuzzy Expert Systems and Fuzzy Reasoning* (Wiley, Hoboken, 2005)

J.J. Schneider, S. Kirkpatrick, *Stochastic Optimization* (Springer, Berlin, 2006)

M. Tajine, D. Elizondo, New methods for testing linear separability. Neurocomputing **47**, 161–188 (2002). doi:10.1016/S0925-2312(01)00587-2

D.A. Van Veldhuizen, G.B. Lamont, Multiobjective evolutionary algorithms: analyzing the state-of-the-art. Evol. Comput. **8**, 125–147 (2000). doi:10.1162/106365600568158

N. Zhang, An online gradient method with momentumnext term for two-layer feedforward neural networks. Appl. Math. Comput. **212**, 488–498 (2009). doi:10.1016/j.camwa.2011.09.028

Y. Zhang, T. Chai, H. Wang, A nonlinear control method based on ANFIS and multiple models for a class of SISO nonlinear systems and its application. IEEE T Neural Netw. **22**, 1783–1795 (2011). doi:10.1109/TNN.2011.2166561

E. Zitzler, M. Laumanns, E. Bleuler, A tutorial on evolutionary multiobjective optimization, in *Metaheuristics for Multiobjective Optimization*, ed. by X. Gandibleux, M. Sevaux, K. Sörensen, V. T'kindt (Springer, Berlin, 2004)

Chapter 4
Case of Study

Abstract This chapter presents a case of study exemplifying the combined application of the finite element method (FEM) and artificial intelligence (AI) based tool in modeling and optimization of manufacturing process. In the study, modeling of an orthogonal cutting process of AISI 1045 steel is carry out by using the FEM. Outcomes of the model, in terms of cutting forces and tool wear are related to the experimental factors (feed, velocity and rake angle) through neural networks models. Finally, a multi-objective optimization process is defined and executed in order to obtain the most convenient factors for this specific process, on different workshop conditions.

4.1 Case Description

In order to exemplify the use of hybrid approach in manufacturing processes, a turning process optimization will be carried out. The objective is to obtain the cutting parameters (cutting speed and feed) and rake angle for guarantying the most convenient values of productivity and tool waste. For achieving this goal, the models relating tool life and cutting forces with the before mentioned independent variables must be obtained for carrying out a multi-objective optimization process.

4.2 Finite Element Method Based Modeling

4.2.1 Model Description

For the first step, i.e. for obtaining the models, the turning process was idealized as an orthogonal cutting process with a depth of cut $a_P = 1.0$ mm. A three level full factorial experimental design was used to simulate the behavior of the cutting

R. Quiza et al., *Hybrid Modeling and Optimization of Manufacturing*,
SpringerBriefs in Computational Mechanics,
DOI: 10.1007/978-3-642-28085-6_4, © The Author(s) 2012

Table 4.1 Experimental
factors and their
corresponding levels

Factors	Levels		
	Low	Medium	High
Cutting speed, v [m/min]	80	140	200
Feed, f [mm/rev]	0.1	0.3	0.5
Rake angle, γ [°]	0	5	10

forces and tool wear, under different values of the experimental factors. In Table 4.1 the selected levels for each factor are shown.

The used workpiece material was AISI 1045 steel. The material behavior was considered as rigid-thermo-viscoplastic, and its hardening was described by the Jonhson-Cook model (Jasper and Dautzenberg 2002):

$$\sigma_{eq} = \left[553.1 + 600.8(\varepsilon_{eq}^P)^{0.234}\right]\left[1 + 0.0134\ln\left(\frac{\dot{\varepsilon}_{eq}^P}{10^{-3}}\right)\right]\left[1 - \left(\frac{T-293}{1440}\right)\right]; \quad (4.1)$$

where σ_{eq} is the Von Misses equivalent stress; ε_{eq}^P, the equivalent plastic strain; $\dot{\varepsilon}_{eq}^P$, the equivalent plastic strain rate and T, the absolute temperature. Values of thermal conductivity, κ; heat capacity, c and emissivity are shown in Fig. 4.1 (Iqbal et al. 2007). The emissivity of the workpiece material was taken as 0.75.

Tool material was uncoated carbide ISO P20. It was considered as perfectly rigid and the temperature dependence of its thermal properties is represented in Fig. 4.2. The emissivity was considered as 0.60.

In Fig. 4.3 the principal geometric dimensions of the model are summarized. The considered section of the tool tip is ten times larger than the feed, f, in every experimental point. The clearance angle had a constant value of 10°, while the rake angle, γ, being an experimental factor changes in each case. The edge radius of the tool is 0.06 mm.

The workpiece height and length were 10 and 50 times larger than feed. This warranties that the boundaries of the model are far enough from the cutting area. The cutting process was analyzed during the time corresponding to the displacement of the tool through the 60% (30 times the feed) of the workpiece length.

At the tool-chip interface a constant friction factor of 0.6 and a heat transfer coefficient of 100 kW/(m² °C) were considered. For free surfaces (in the workpiece and tool) a heat transfer coefficient with the environment of 20 kW/(m² °C) was selected.

Mesh size was selected according of the feed, having approximately 700 elements the mesh in the workpiece and 1500 in the tool.

4.2.2 Outcomes of the FEM

The simulation process was implemented through a ALE approach which allowed to obtain the chip formation process. In Fig. 4.4 this process for medium-level

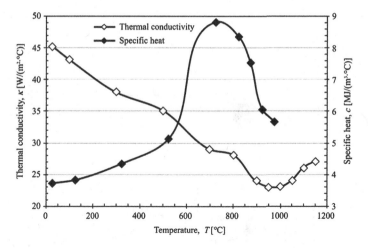

Fig. 4.1 Thermal properties of the workpiece material

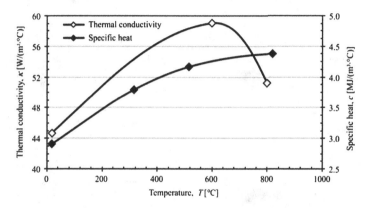

Fig. 4.2 Thermal properties of the tool material

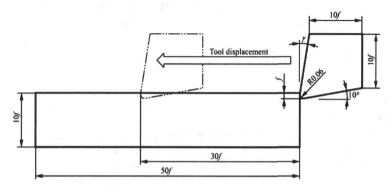

Fig. 4.3 Dimensions of the model

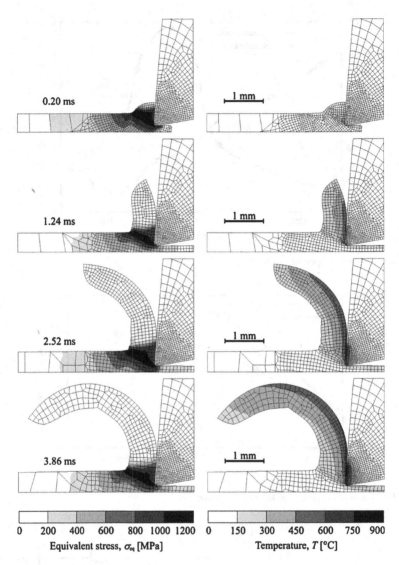

Fig. 4.4 Simulation of the chip formation processes for the medium-level values of the experimental factors

values of the experimental factors ($v = 140$ m/min, $f = 0.3$ mm/rev, $\gamma = 5°$) is shown:

For these same values of experimental factors, the cutting (tangential) and feed (axial) components of the force where obtained (Fig. 4.5). Also, the maximum temperature in the chip and in the tool where simulated (Fig. 4.6).

In order to estimate the tool life, the wear rate in the rake surface of the tool is computed from values of pressure, sliding velocity and temperature, which are

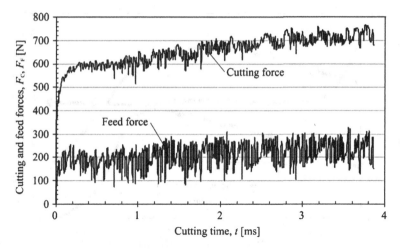

Fig. 4.5 Simulated forces for medium-level values of experimental factors

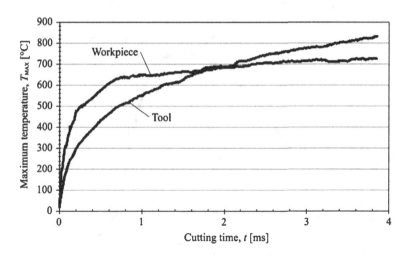

Fig. 4.6 Simulated temperature for medium-level values of experimental factors

evaluated for different points on the tool-chip interface. The corresponding outcomes, for the medium-level experimental factors are shown in Fig. 4.7.

By using the Usui's expression (Yen et al. 2004):

$$\frac{dw}{dt} = Ap_n v_s \exp\left(-\frac{B}{T}\right);$$ (4.1a)

where, for plain carbon steels versus uncoated carbide tools:

$$\begin{aligned} A &= 7.80 \times 10^{-9} \mathrm{m}^2/\mathrm{MN}, \ B = 5.302 \times 10^3 \mathrm{K}^{-1} \quad \text{if } T < 1150 \text{ K} \\ A &= 1.198 \times 10^{-2} \mathrm{m}^2/\mathrm{MN}, \ B = 2.195 \times 10^4 \mathrm{K}^{-1} \quad \text{if } T \geq 1150 \text{ K} \end{aligned}\right\};$$ (4.1b)

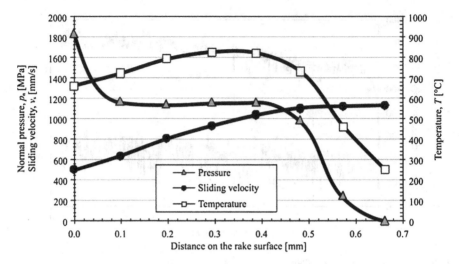

Fig. 4.7 Parameters at the rake face for medium-level values of experimental factors

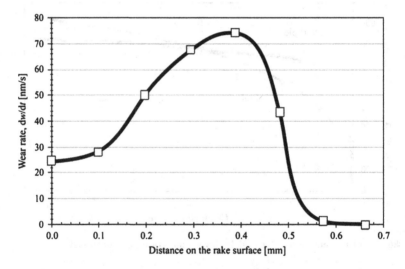

Fig. 4.8 Wear rates at the rake surface for the for medium-level values of experimental factors

the wear rate is evaluated for each point, which allows constructing the graph of wear rate on the rake surface (Fig. 4.8).

As can be seen, the maximum wear rate corresponds to a value of 72.7 nm/s located at 0.39 mm from the beginning of the rake surface.

This process was applied to each point in the experimental design, obtaining the corresponding values of cutting force, temperature, normal pressure and sliding velocity at the rake face. With the last three parameters, the wear rate is computed for each experimental point. The obtained values are shown in Table 4.2.

Table 4.2 Simulated outcomes of the cutting processes

V [m/min]	f [mm/rev]	γ [°]	F_C [N]	T [°C]	p_n [MPa]	v_s [mm/s]	dw/dt [nm/s]
80	0.1	0	268	397	1060	397	1.20
80	0.1	5	251	412	1230	475	1.98
80	0.1	10	234	434	1530	638	4.21
80	0.3	0	744	624	1050	480	10.6
80	0.3	5	712	641	1120	560	14.8
80	0.3	10	679	655	1240	670	21.4
80	0.5	0	1220	784	1040	571	30.7
80	0.5	5	1170	798	1080	623	37.2
80	0.5	10	1160	807	1120	697	44.9
140	0.1	0	290	593	1130	950	18.4
140	0.1	5	272	620	1230	1060	26.8
140	0.1	10	255	645	1350	1160	38.0
140	0.3	0	813	928	1150	1100	175
140	0.3	5	726	823	1190	988	72.7
140	0.3	10	742	987	1260	1360	556
140	0.5	0	1320	1140	1170	1280	3220
140	0.5	5	1300	1180	1200	1440	5690
140	0.5	10	1240	1210	1220	1590	8670
200	0.1	0	262	606	1020	1090	20.8
200	0.1	5	246	634	1060	1210	28.9
200	0.1	10	231	658	1090	1230	35.2
200	0.3	0	737	946	1050	1240	236
200	0.3	5	718	982	1050	1550	494
200	0.3	10	672	1010	1100	1540	754
200	0.5	0	1210	1160	1070	1420	4050
200	0.5	5	1160	1200	1090	1620	7141
200	0.5	10	1100	1240	1100	1830	12100

4.3 Empirical Modeling

4.3.1 Statistical Modeling

For establishing the mathematical relationship between the studied variables (cutting force and wear rate) with the experimental factors (cutting speed, feed and rake angle), the first choice are the statistical regressions because they are simple and have a solid mathematical foundation.

By applying the multiple regression technique, the following model was obtained for the cutting force:

$$F_c = 2421 f^{0.9613} (\cos \gamma)^{5.763}. \tag{4.2}$$

with a R-squared of 0.994, which means that the fitted model explains more than 99% of the variability in the dependent variable. The ANOVA (Table 4.3) shows that there is a statistically significant relationship between the cutting force and the

Table 4.3 ANOVA for the statistical model of the cutting force

Source	Sum of squares	Degrees of freedom	Mean square	F-ratio	p-value
Model	4.099×10^6	3	1.366×10^6	803.4	0.0000
Residual	0.039×10^6	23	0.002×10^6		
Total	4.138×10^6	26			

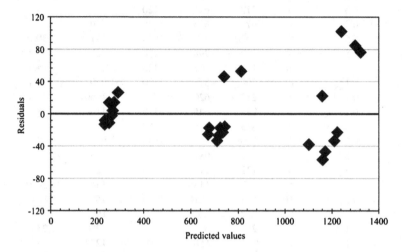

Fig. 4.9 Residual distributions for the statistical model of the cutting force

two included independent variables. The cutting velocity was removed from the model because the t-Student test showed that this term was not significant at the 90% or higher confidence level.

Figure 4.9 shows that residuals follow a normal distribution without any recognizable trend. Taking into account all the before mentioned aspects the obtained statistical regression can be accepted as a proper model for the cutting force in this case of study.

For the wear rate, several multiple regressions were tried, being the most convenient than that represented by the following equation:

$$\frac{dw}{dt} = 3.694 \times 10^{-6} \frac{V^{4.240} f^{2.710}}{(\cos \gamma)^{56.35}}; \qquad (4.3)$$

which has an R-squared of 0.869, indicating that the model as fitted explains only an 86.9% of the variability of the independent variable.

4.3.2 Neural Network-Based Modeling

In order to obtain a more accurate model for the wear rate, a multilayer perceptron was selected to be trained with the data. For training the network, not only the

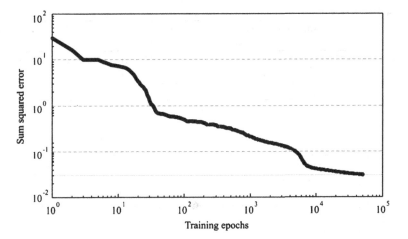

Fig. 4.10 Training process of the neural network for the wear rate model

input data but also the output were normalized in the interval [0, 1] by using a linear interpolation. The output, also, was logarithmized in order to counteract the high difference between higher and lower values.

A two-layers Multi-layer preceptron (MLP) was designed with three neuron and sigmoid activation function in the hidden layer and one neuron and linear activation function in the output layer. The network has 16 free parameters (weights and biases), so the data is enough for training the network.

The training process was carried out by using a back-propagation algorithm with adaptive learning rate and momentum. An initial learning rate of 0.01 and a momentum constant of 0.9 were established for the training process. The prescribed sum squared error of 0.03 was reached after 49812 epochs (see Fig. 4.10).

The obtained model can be represented by the following MATLAB function.

```
function W = wearrate(v, f, g);
W1 = [    −0.1558,      −1.6702,      −0.5204; ...

               2.5610,       5.7694,      −5.9064; ...
              −9.0230,       0.9992,       0.1167   ];
B1 = [     1.4947;      −2.5593;      −2.7090   ];
W2 = [    −1.4144,       0.0894,      −3.0133   ];
B2 =        1.3793;
X = [    (v − 80)./(200 − 80); ...

             (f − 0.1)./(0.5 − 0.1); ...
             (g − 0)./(10 − 0)         ];

y = W2*logsig(W1*X + B1) + B2;
W = exp(0.1823 + (9.3985 − 0.1823).*y);
```

Table 4.4 ANOVA for the neural model of the wear rate

Source	Sum of squares	Degrees of freedom	Mean square	F-ratio	p-value
Model	3.1286×10^8	21	1.4898×10^7	19.1	0.0000
Residual	0.0391×10^8	5	0.0781×10^7		
Total	3.1677×10^8	26			

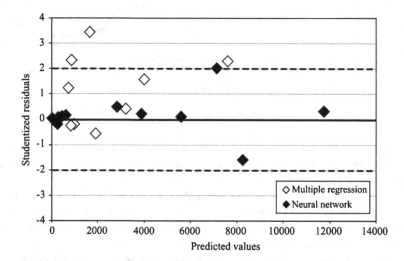

Fig. 4.11 Residuals of the models of wear rate

It has an R-squared statistic equal to 0.985, which means that the fitted model explains more than 98% of the variability of the dependent variable. From the ANOVA (Table 4.4) it can be seen that there is a statistically significant relation between the variables with more than 99% of confidence level.

A residual plot was carried out in order to compare the behavior of the neural model with that of the previously fitted multiple regression. As it is shown (Fig. 4.11) the residuals from the neural model are notably lower and none of them is outside of the interval [−2, 2], indicating that there are not outliers in the residuals. On the contrary, in residuals of the regression, three values are outside of the interval [−2, 2] and one of them is outside of [−3, 3].

Taking into account these issues, it can be considered that the neural network-based model is more accurate than the multiple regression. Therefore, it was selected to be used in the optimization of the process. In Fig. 4.12 the neural model is graphically represented.

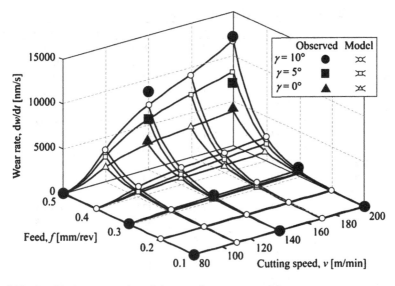

Fig. 4.12 Graphical representation of the neural wear rate model

4.4 Optimization

The decision variables involved in the optimization process were the same experimental factors: cutting speed, v; feed, f; and rake angle, γ. The optimization targets were the material removal rate:

$$MRR = 1000vfa_\mathrm{P}; \tag{4.4}$$

which defines the productivity of the process, and the wear rate, given by the previously obtained neural model:

$$\frac{\mathrm{d}w}{\mathrm{d}t} = f(v, f, \gamma); \tag{4.5}$$

which represents the tool waste. Evidently, the material removal rate must be maximized while the wear rate must be minimized. As both objectives are mutually conflictive, the optimization was multi-objective, so an a posteriori approach was selected.

Only one constraint was considered: the cutting power, which can be computed as:

$$P_\mathrm{c} = \frac{vF_\mathrm{c}}{6 \times 10^4}; \tag{4.6}$$

where F_c is the cutting force that can be computed by the Eq. (4.2).

Table 4.5 Non-dominated solutions obtained after the optimization

No.	v [m/min]	f [mm/rev]	g [°]	dw/dt [nm/s]	MRR [mm³/min]
1	200	0.48	4.0	5815	96496
2	199	0.46	2.5	3411	91757
3	194	0.45	1.4	2362	86925
4	200	0.39	1.9	1147	77668
5	199	0.37	1.0	757	73440
6	197	0.36	1.0	636	70269
7	199	0.33	1.4	486	66333
8	190	0.33	1.0	435	63134
9	200	0.27	4.6	225	53181
10	190	0.25	1.2	142	47870
11	190	0.23	6.9	158	44452
12	181	0.22	7.2	134	40300
13	188	0.19	1.4	55	34973
14	81	0.37	4.6	23	30434
15	81	0.34	1.9	15	27582
16	81	0.30	5.0	11	23914
17	81	0.27	2.1	8	21659
18	81	0.19	0.3	4	15625
19	81	0.14	2.7	3	11730

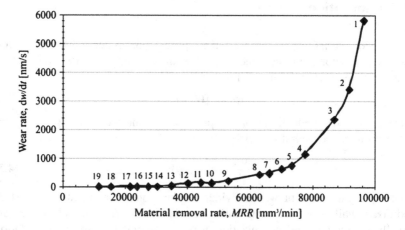

Fig. 4.13 Pareto front

For carrying out the optimization process a micro genetic algorithm (Quiza et al. 2006) was used. Thousand and 25 individuals were used in the static and dynamic population populations, respectively. The mutation likelihood value was $10^{-4} \times 100$ evolutionary periods, each one consisting in 50 epochs, were executed. After that, the non-dominate solutions shown in Table 4.5 were obtained.

From these values, which can be graphically represented in a Pareto front (see Fig. 4.13), the most convenient cutting parameters can be selected depending on the specific workshop conditions.

For example, if the most productive process is desired, the cutting parameters corresponding to point 1 must be selected. There, the highest productivity but also the highest tool wear are achieved. On the contrary, if used cutting tools are expensive and time is not an important factor, cutting parameters corresponding to point 19 are the most convenient, allowing obtaining low tool wear although at a low productivity. Finally, some other combinations can be used, depending on the specific economic conditions of the workshop.

4.5 Concluding Remarks

As can be shown, combining AI and FEM can result in a powerful and versatile tool for modeling and optimizing manufacturing processes. The capability of the FEM for obtaining approximate solutions of complex differential equations system, together with the advantages of the AI-based techniques for identifying complex nonlinear patterns, makes this combination very effective.

Nevertheless, the shortcomings and disadvantages of both methods cannot be ignorned. For example, FEM gives only approximate solutions, which depend on the quality of the mesh or the numerical techniques used for solving algebraic systems of equations. On the other hand, AI modeling methods are purely empiric; therefore, they cannot supply knowledge about the phenomena they are describing.

Is spite of the notable successes attained by the hybrid approach in manufacturing processes modeling and optimization, more research is needed in this field in order to enhance the capability not only of both involved tools (AI and FEM) but also of the way they are linked. However, with the constant increase in the computation power of the computer machines, in last time, it can be expected that hybrid approaches, in a near future, can be widely used not only in academic research but also in the workshop environment.

References

S.A. Iqbal, P.T. Mativenga, M.A. Sheikh, Characterization of machining of AISI 1045 steel over a wide range of cutting speeds. Part 2: evaluation of flow stress models and interface friction distribution schemes. Proc. Inst. Mech. Eng. B-J Eng. **221**, 917–926 (2007). doi:10.1243/09544054JEM797

S.P.F.C. Jasper, J.H. Dautzenberg, Material behaviour in conditions similar to metal cutting: flow stress in the primary shear zone. J. Mater. Process. Tech. **122**, 322–330 (2002). doi:10.1016/S0924-0136(01)01228-6

R. Quiza, M. Rivas, E. Alfonso, Genetic algorithm-based multi-objective optimization of cutting parameters in turning processes. Eng. Appl. Artif. Intell. **19**, 127–133 (2006)

Y.-C. Yen, J. Söhner, B. Lilly, T. Altan, Estimation of tool wear in orthogonal cutting using the finite element analysis. J. Mater. Process. Tech. **146**, 82–91 (2004). doi:10.1016/S0924-0136(03)00847-1

Index

A

A posteriori approach (optimization), 65
A priori approach (optimization), 65
ACO. *See* Ant colony optimization, 71
Adaptive neuro-fuzzy inference system, 62
AI/FEM hybrid models, 5
AISI 1045 steel, 80
ALE formulation. *See* Arbitrary Lagrangian-Eulerian formulation, 30
ANFIS. *See* Adaptive neuro-fuzzy inference system, 62
ANFIS architecture, 62
ANFIS training, 64
Ant colony optimization, 71
Arbitrary Lagrangian-Eulerian formulation, 30
ART. *See* Adaptive resonance theory, 51
ART-1, 51
ART-1 topology, 53
Artificial intelligence, 39
Artificial neural networks. *See* Neural networks, 3, 40, 61

B

Biological neuron, 40
Brozzo's damage criterion, 34

C

Case of study, 79
Chip formation, 80
Clustering algorithm, 51
Cockcroft-Latham damage criterion, 34

Cohesive friction model, 33
Coulomb's simple friction model, 31
Crossover, 68

D

Design of experiments, 79
Differential equation set, 13

E

Elastic strain component, 21
Elastic-linear work-hardening behavior, 22
Elastic-nonlinear work-hardening behavior, 22
Elastic-perfectly plastic behavior, 22
Elastic-plastic matrix, 29
Empirical models, 2
Error back-propagation algorithm, 44
ES. *See* Evolution strategies, 66
Eulerian formulation
    boundary conditions, 26
    constitutive relation, 25
    definition, 24
    finite element formulation, 29, 30
    initial conditions, 27
    kinematic relation, 25
    motion equation, 26
    rigid-plastic behavior, 27
Evolution strategies, 66
Evolutionary computation, 66
Evolutionary multi-objective optimization, 68

R. Quiza et al., *Hybrid Modeling and Optimization of Manufacturing*,
SpringerBriefs in Computational Mechanics
DOI: 10.1007/978-3-642-28085-6, © The Author(s) 2012

E (*cont.*)
Exact RBFN, 48
Experimental stress-strain curve, 20

F
FEM. *See* Finite element method, 13
FEM model settings, 80
FEM/AI hybrid models, 4
Finite element, 14
Finite element method, 2, 13
Fourier's law, 30
Freudenthal's damage criterion, 34
Fuzzy FEM, 8
Fuzzy inference rules, 59
Fuzzy inference system, 60
Fuzzy logic, 58
Fuzzy operations, 59

G
GA. *See* Genetic algorithms, 66, 67
Galerkin method, 15
Generalized midpoint rule, 30
Genetic algorithms, 66

H
Hamilton's principle, 17
Heat generation, 31
Hebbian rule, 50
Hillerborg's fracture energy, 36
Hollomon's stress-strain law, 22
Hooke's law, 21
Hopfield networks, 49
Hybrid FEM/AI optimization, 7

I
Initial yielding criterion, 24
ISO P20 tungsten carbide, 80

J
J-C law. *See* Johnson-Cook's
          model, 23
Johnson-Cook's shear failure
          model, 35
Jonhson-Cook's model, 23

K
Kohonen networks. *See*
          Self-organizing maps, 54

L
Lagrangian formulation
    boundary conditions, 29
    constitutive relation, 28
    definition, 27
    finite element formulation, 29, 30
    initial conditions, 29
    kinematic relation, 27
    motion equation, 29
Linear elastostatics, 16
Ludwik's stress-strain law, 22
Mamdani fuzzy inference system, 60
Material hardening function, 22, 26
MATLAB, 87
McCulloch and Pitts neuron model, 40
Membership function, 58
Modified Coulomb's friction model, 32
Multilayer perceptron, 43
Multi-layer preceptron, 87
Multi-objective optimization, 64
Multiple regression, 85
Mutation, 68

N
Neural networks
    activation function, 41
    definition, 40
    feed-forward, 42
    ggregation function, 41
    layer, 41
    learning procedures, 43
    mathematical indetermination, 57
    misusing, 56
    over-fitting, 58
    recurrent, 42
    statistic tests, 58
    synapse, 40
    topology, 41
Neuro-fuzzy system, 61
Newton's law of cooling, 31
Nodal force vector, 19
Nodes, 18
Numerical integration, 30

O
Optimization, 3, 64, 89
Oxley's thermo-viscoplastic model, 23

P
Paretian dominance, 66
Pareto front, 66, 90

Pareto optimal, 66
Particle swarm optimization, 74
Perceptron rule for Hopfield networks, 50
Phenomenological models, 2
Plastic strain component, 21
Plasticity, 20
Power law of thermo-viscoplasticity, 23
Progessive approach (optimization), 65
PSO. *See* Particle swarm optimization, 71, 74

R
Radial basis function networks, 46
  learning, 47
Rate-independent plasticity, 20
RBFN. *See* Radial basis function network, 41,
  42, 46

S
SA. *See* Simulated annealing, 3, 71, 74
Selection, 67
Self-organizing maps, 54
Shape functions, 19
Simulated annealing, 74
Single-objective optimization, 64
Soft computing, 39
Specific distortion damage criterion, 34
Stiffness matrix, 19, 29
Strain rate-independent plastic models, 22
Strain-displacement matrix, 19
Stress index parameter damage criterion, 34
Sugeno fuzzy inference system, 60
Swarm intelligence, 71
Swift's stress-strain law, 22

T
Thermal properties, 80
Thermo-viscoplastic models, 23
Tresca's yielding criterion, 24

U
Usui and Shirakashi's friction model, 32
Usui's wear model, 83

V
Variational method, 15
Viscoplastic models, 23
Viscoplasticity, 20
Voce's stress-strain law, 22
Von Mises' yielding criterion, 24

W
Weak form, 14
Weighted residual method, 15

Y
Young's modulus, 21

Z
Zerilli and Armstrong constitutive
  model, 23